GENETICS IN MINUTES

TOM JACKSON

GENETICS IN MINUTES

TOM JACKSON

Quercus

CONTENTS

Introduction

In its simplest terms, genetics is the study of inheritance. However, looking a little deeper, there is nothing simple about it. Genetics tells us how a body can grow from a single cell; it shows how life on Earth has changed in a myriad ways over billions of years; and it forms a central plank in the fight against disease. What's more, it also has the potential to create new technology that will transform society, ensuring health for all and perhaps even allowing us to control the future development of our species and reshape the living world.

As a science, genetics is relatively new: its foundations date from the 1850s, but those many different strands were not drawn into a single field until the early 20th century. It was slow going at first, and not until the 1950s did the great mysteries of genetics begin to give up their meanings. First was the discovery of the DNA double helix, and after that the so-called 'Central Dogma', which shows how an inanimate chemical code can result in a living body. Progress accelerated rapidly as we unlocked more of the secrets of the gene, but even today, despite huge advances, there are many riddles within our DNA that we are still to solve. We may have learned how to decipher the genetic code, but the work of translating what it all means is still proceeding.

Genetics draws from many fields, such as chemistry, biology, agriculture, engineering, even information theory and statistics. For many, the expectation is that genetics can tell us exactly who we are, what's 'in the genes'. Long before the science of genetics existed, our ancestors would have understood that a child was a unique blend of characteristics inherited from its parents. However, the extent to which the nature of our genetic code rules our behaviours and personalities is proving the most difficult puzzle to solve. Perhaps the latest interests of genetics, such as stem cell research, epigenetics and artificial biology, will provide those missing pieces – certainly these intriguing areas of research suggest that genetics will continue to have a huge influence on medicine and our understanding of what it means to be human in the 21st century and beyond.

Life

What is life? In a nutshell, scientists would define it as a self-replicating process that requires at least one 'thermodynamic cycle'. To put that another way, something that is alive is able to make a copy of itself, and it does this by harnessing a source of energy, using it to transform chemical resources in some way. The supply of energy must be continuous; if the energy source were to become unavailable, or the life form became unable to tap it, then the result would be death. That is something else unique that life can do: it can die.

According to this definition, the simplest life form is a strand of nucleic acid, something like RNA (see page 100). This chemical is able to use its own molecule as a template for a copy of itself. However, such a life is incredibly precarious, and over billions of years of evolution, a multitude of life forms have developed abilities that ensure survival. These abilities are set out in genes, and they govern the success or failure of a life. To understand life, one must begin with genetics.

Types of organism

The number of different types, or species, of organism on Earth is estimated to be anywhere between 3 and 30 million, with most biologists erring towards about 9 million.

The simplest and oldest life forms are the bacteria, which have a body made from a single tiny 'prokaryotic' cell (see page 58). They are joined by the archaea, which to the uninitiated look more or less the same but have some important distinctions. Other single-celled organisms, including things like amoebae and protozoa, have much larger and more complex cells, and this 'eukaryotic' cell type (see page 60) is the one used by multicellular organisms such as plants, animals and fungi.

Every species of organism has a unique way of life, but members of any biological group share more characteristics with each other than with the members of other groups. However, all life forms share a set of abilities: they sense the surroundings, excrete, reproduce, grow, respire and require nutrition.

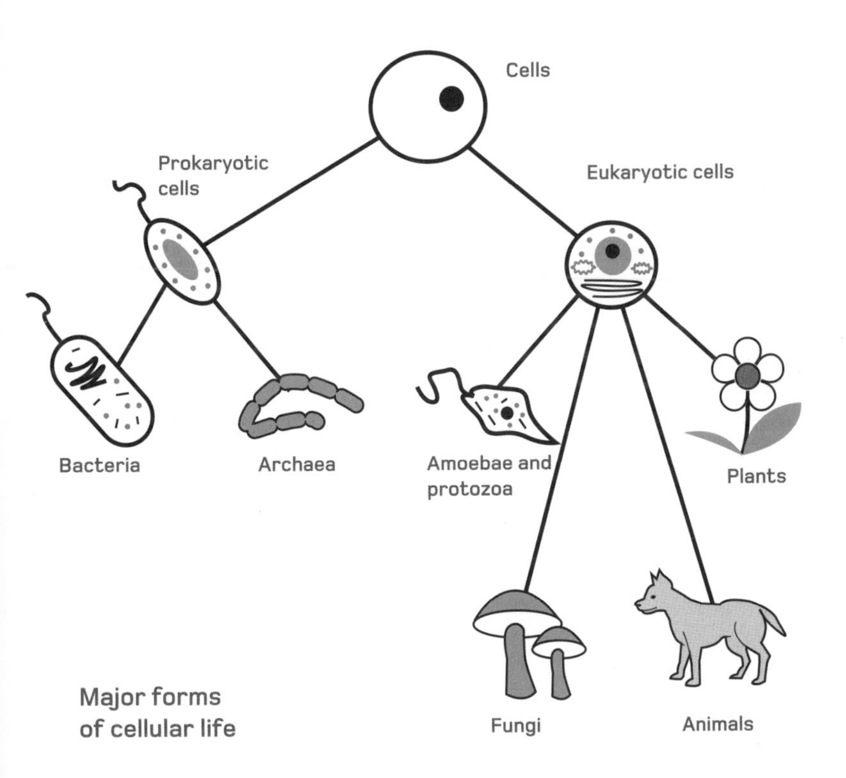

Major forms of cellular life

Metabolism

The broad term 'metabolism' encapsulates the link between chemical activity and life – the myriad chemical processes that are occurring inside every organism are described as its metabolism. In very general terms, these include the way the organism handles its energy supply, and how it uses this to grow and repair its body, making use of simple chemical building blocks.

Metabolic processes fall into two general types: anabolism and catabolism. The former involve building larger, more complex and more ordered structures out of smaller units. (That is why a sports cheat might use an 'anabolic steroid', a chemical that builds muscle.) Catabolism, in contrast, involves splitting large structures into smaller ones (this includes processing unwanted waste materials to generate energy). Anabolic and catabolic processes are constantly working together to release manageable packets of energy and then put them to work in keeping the organism alive.

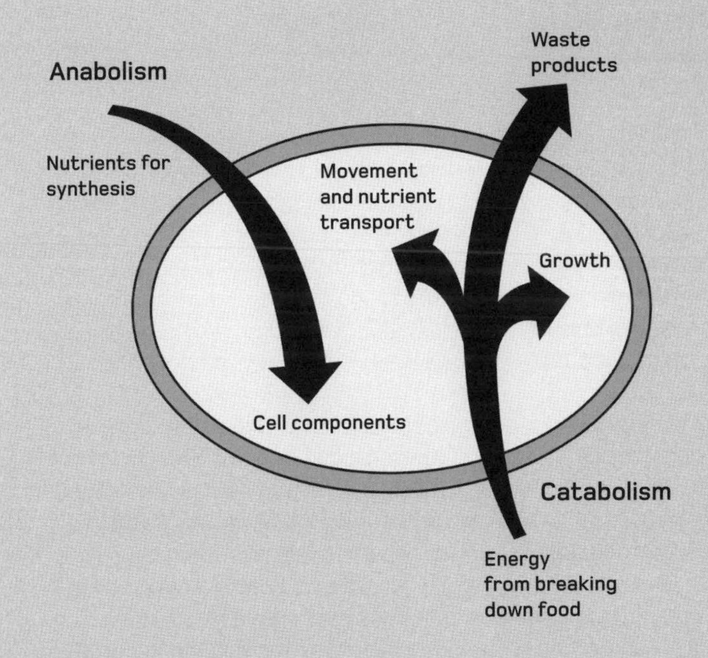

Basic metabolic processes

Feeding

Every living thing must feed, or putting it more precisely, they must access a source of nutrition. Plants get this in the form of sugars from photosynthesis and mineral nutrients absorbed from their surroundings (soil is a good place to start). Animals and fungi get their nutrition from the bodies of other organisms. Some single-celled organisms can get nutrition using both techniques!

Nutrition has two main purposes. First, it is a source of chemical energy that can be extracted and put to work in the body (the best examples of this are glucose and other sugars). The second purpose is as a stockpile of the raw ingredients required to build a body. The requirements of different organisms vary wildly: plants are able to build everything they need from water, carbon dioxide, and a menu of minerals such as nitrates and phosphates, while animals need more complex nutrition, such as fats, starches, proteins and a range of crucial helper chemicals, known collectively as 'vitamins'.

Respiration

When most people hear the term 'respiration', they tend to assume it relates to breathing. But while this is indeed the word's common medical context, biology gives it a wider meaning: in fact, all organisms respire, whether or not they breathe in and out in the way that vertebrate animals do.

Biologically, respiration is defined as the metabolic process that releases energy from sugar or other chemical fuels. Typically, this involves the fuel molecules being oxidized – exactly the same chemical reaction involved when materials combust in air. The respiration of glucose, one of the most common sugars, for example, can be written in the form of a chemical equation as shown opposite. This demonstrates that glucose reacts with oxygen to produce carbon dioxide and water, plus some energy. If raw glucose is burnt in air, the reaction produces flames and heat, but within a living cell it can be heavily regulated, allowing small packets of energy to be released in several steps.

$$C_6H_{12}O_6 + 6O_2 \rightarrow 6CO_2 + 6H_2O + \text{energy}$$

glucose + oxygen ⟶ carbon dioxide + water + energy

Photosynthesis

As the word suggests, 'photosynthesis' is the process of 'making with light', and the end product in question is glucose sugar. Photosynthesis takes place in the leaves and the other green parts of plants and other photosynthetic organisms. The colour is important because the energy from sunlight is absorbed by a pigment chemical called chlorophyll in the plant's cells – chlorophyll itself appears green because it traps the blue and red wavelengths of sunlight while reflecting other colours.

Chemically, photosynthesis is the reverse of respiration, with carbon dioxide and water molecules being combined to make glucose molecules and oxygen, all powered by the energy channelled from the chlorophyll molecules. While carbon dioxide is the waste product of respiration, photosynthetic organisms produce waste oxygen, which is released into the air. Nearly all of the oxygen in Earth's atmosphere (about 20 per cent of all the air) originated as the by-product of photosynthesis.

$$6CO_2 + 6H_2O \xrightarrow[\text{chlorophyll}]{\text{light energy}} C_6H_{12}O_6 + 6O_2$$

$$\text{carbon dioxide + water} \xrightarrow[\text{chlorophyll}]{\text{light energy}} \text{carbohydrates + oxygen}$$

Growth

To qualify as living, an organism needs to be able to grow – at least at some point in its life. For most complex organisms this is a simple thing to verify. The majority of multicellular life forms – those with bodies of more than one cell – grow from a single cell into an embryo and on to a fully developed adult. This growth is achieved by the division of cells (see pages 76 and 132), with every cell in the body being descended by some route from that first single cell, known as the zygote.

The growth of single-celled organisms – things like bacteria and amoebae – is less clear cut. They too can divide their cells, but instead of creating a larger body, they produce a new and independent individual. In these cases, growth and reproduction are two sides of the same coin. Therefore, the best definition of growth is the ability to produce new cells from older cells. This is the concept that lies at the heart of cell theory (see page 72), a central tenet of life science.

Reproduction

It could be said that the primary goal of an organism is to survive. However, that survival is really a means to an end – all organisms are striving to make a copy of themselves or something close to it. In other words the true purpose of biological life is reproduction. There are many modes of reproduction, ranging from organisms simply dividing in two to a complex process of courtship, mate selection and parental care. However, broadly speaking there are two types of reproduction: sexual and asexual. The former involves two parents and the latter requires only one (see pages 118 and 120).

The struggle to survive and reproduce is the driving force behind evolution by natural selection (see page 184), the process that shapes the millions of species that live on Earth. However, this evolution is a by-product of reproduction. The genetic purpose of reproduction is to make new copies, and many of them, of the DNA molecules in all bodies, reproducing the information that we call genes.

Excretion

Just as an organism takes in nutrients and other raw materials from its surroundings, it must also remove the waste products of metabolism – a process known as excretion. Despite common usage, the voiding of the bowel, passing faecal matter, out of the body is not actually excretion in biological terms: instead, it is defecation or egestion. The crucial difference is that the unused food has not really entered the body – it has only passed through the gut, a hollow tube that runs *through* the body. True excretion is the process of taking waste products – which may be harmful if left to accrue – from the body's tissues and expelling them.

In human biology the chief mode of excretion is urination, whereby excess water and nitrogen-rich waste in the form of urea are released. Excretion can also occur directly through the skin as sweating. In addition, the release of carbon dioxide generated by respiration processes is also a form of excretion.

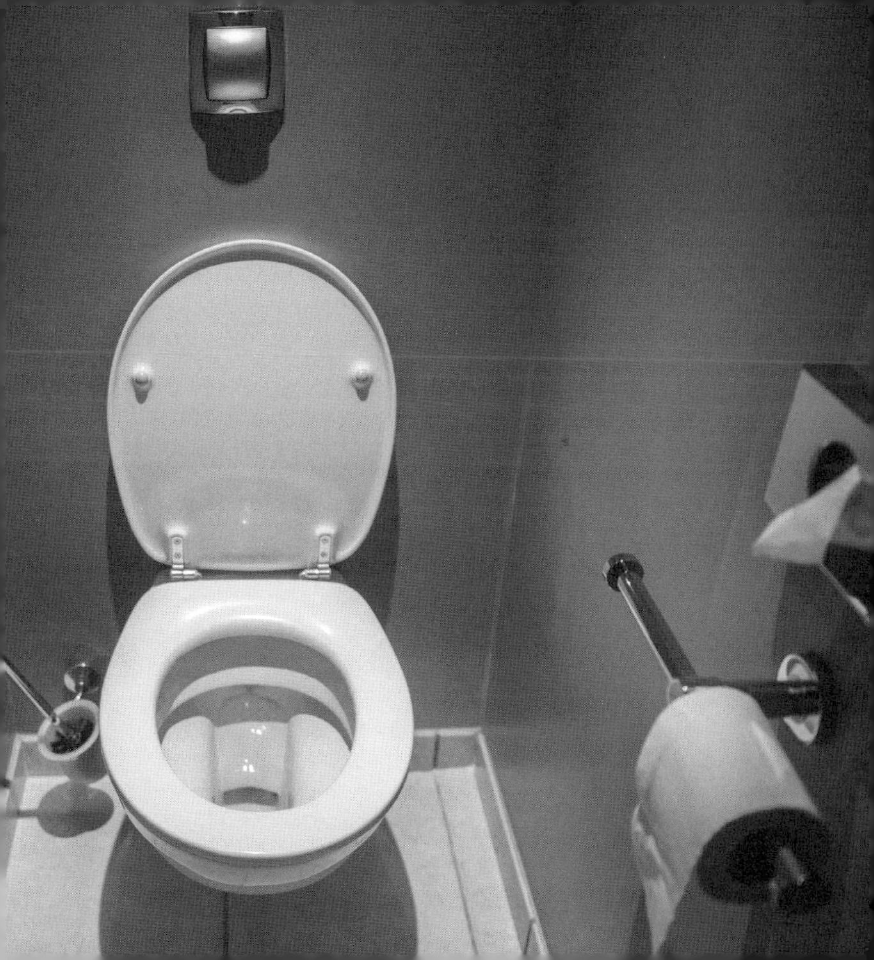

Senses

All life forms are able to detect changes in their surroundings and respond to them. For single-celled organisms this may be simply a matter of detecting a chemical change, such as the salinity of water or the presence of nutrients or toxins. Plants, meanwhile, are sensitive to light, gravity and sometimes pressure – they grow towards light and away from the pull of gravity, and some adjust their growth patterns to wrap themselves around other objects they contact.

Animal senses are much more advanced, befitting their active lifestyles. The five used by humans are somewhat ubiquitous: hearing, smell, taste, vision and touch. The last of these is a complex mix of detectors on the body surface, sensitive to heat, cold, vibrations and pressure. Other animals can sense things beyond a human's abilities. Many insects and other arthropods can detect ultraviolet light; sharks and their cousins can detect electrical activity in another body, while many other animals appear to sense Earth's magnetic field.

Inheritance

The science of genetics is relatively new. Its first steps were made in the 1850s and the term 'genetics' was not coined until 1905. It was, however, a new word for an old field of enquiry: inheritance. Since prehistoric times it was well understood that children inherited some of the attributes of their parents. Characteristics such as hair colour, face shape and height are passed on in families, from generation to generation. This applies as much to animals and plants – especially those used in farming – as it does to humans.

The search for the mechanisms of inheritance led to the science of genetics and the theory of evolution, but it did not begin there. The ancient Greek theory was 'pangenesis', which proposed that every body part sent information via the semen and menstrual blood to create a tiny person, or *homunculus*, that grew inside the mother. Charles Darwin himself espoused something like this, saying inherited traits travelled between generations as a swarm of tiny packets called 'gemmules'.

The gene

The term 'gene' was coined in 1909 by the Danish botanist Wilhelm Johannsen. Its roots lie in the word 'genesis' meaning origin. Charles Darwin and his colleagues in the late 1800s referred to a still-hypothetical 'genetic' material that transmitted inherited traits. The study of that process became known as genetics in 1905 (thanks to English biologist William Bateson), and soon after Johannsen introduced the concept of the gene.

Johannsen had no idea what form genes took. His term simply meant a unit of inheritance: the genes inherited from the parent carry the instructions required to build the body of a child. The term is also used to describe particular measurable characteristics, so there is a gene for hair type, eye colour, etc. However, today we know that genetic material is a code-carrying molecule of DNA, so a section of DNA can also be described as a gene. Matching this chemical definition of genes with the anatomical one is a key goal of genetic research.

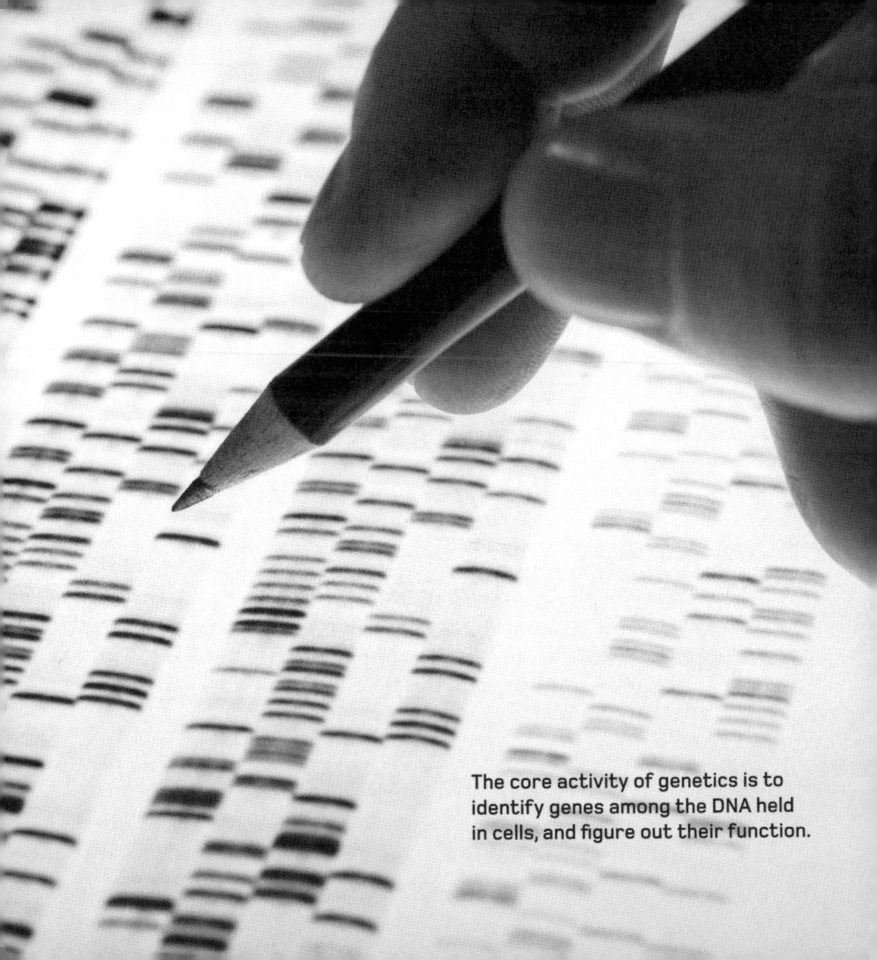

The core activity of genetics is to identify genes among the DNA held in cells, and figure out their function.

Gregor Mendel

Perhaps surprisingly, the founding figure of genetics was a German-speaking monk, living in the northern reaches of the Austro-Hungarian Empire in the mid-19th century. Gregor Mendel's work, carried out in the cloistered garden of the Abbey of St Thomas in Brno (now a Czech city), was completely ignored from its publication in 1866 to the start of the 20th century, but nevertheless it contained the basic tenets of genetics that still apply today.

Mendel (1822–84) made his discoveries through experiments breeding pea plants in his garden. He had no knowledge of DNA, referred little to cell biology and, instead of the term 'gene', used the word 'factor'. However, Mendel was able to glean some universal rules of genetics from the way the different characteristics of the pea plants were passed from generation to generation. These fundamental rules are the foundations of the core inheritance process, which is called Mendelian genetics in his honour.

Mendel's crosses

Gregor Mendel made his discoveries by diligently controlling which pea plants were allowed to breed with which others. He was aided in this endeavour by the fact that peas can self-cross, meaning a plant can use its own pollen to produce seeds.

Mendel identified several inherited traits, such as flower colour or shape and plant height. He worked on all these traits, but taking height as our exemplar, Mendel isolated a tall plant that always produced tall daughter plants when crossed with itself, and a short plant that always produced short offspring. He then cross-pollinated these two plants to produce offspring (seeds) with one tall and one short parent. He found the first generation of offspring grew into tall plants. Next he self-crossed one plant from the new generation. Three quarters of *its* offspring were tall, a quarter were short. The same thing happened for all the traits he tested. Mendel's theories of inheritance were deduced from these startlingly consistent results.

Mendel’s cross-pollination experiment

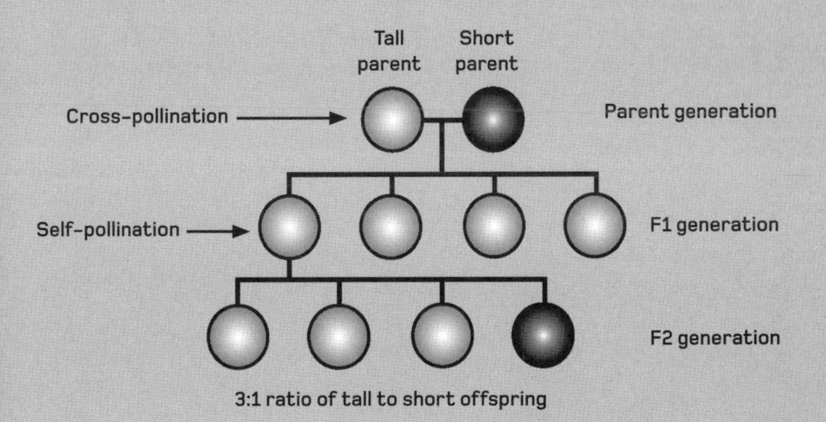

Mendel's Laws

Using the results from his many thousands of breeding experiments carried out over several years, Gregor Mendel outlined what he saw as universal truths about inheritance.

Mendel's 'Law of Segregation' said that each plant had two versions of each factor (gene). When it came to making pollen, the paired versions of each factor were always split. Any offspring would inherit only one version from each parent, with the two combining making a new pair. Another rule, the 'Law of Independent Assortment', states that every factor moves between generations independently of the others. A third law, the 'Law of Dominance' asserts that *some* types of factor have a hierarchy that leads to dominant ones being expressed in the organism's outward appearance, while recessive ones remain hidden. Later research would come to qualify the second law, and some regard the third as less significant because it does not apply to all factors, but together these laws have become the foundation stones of classical genetics.

Mendel's diligent experiments with pea plants gave the first insight into how inherited factors controlled development.

Phenotype

Classical genetics draws a line between our two definitions of a gene: a gene can be understood as a chemical entity – a piece of DNA – or as an inherited trait, anatomical or otherwise. Mendel's discoveries showed that the two concepts were not interchangeable. To illustrate this, geneticists invented the term 'phenotype'.

The phenotype is the outwardly expressed end result of the genes that are inherited. It is the tallness of the pea plant, the colour of your hair or the body plan of an insect. It can also relate to animal behaviours (sometimes referred to as the 'extended phenotype'). There is often a degree of learning involved in behaviours, such as migration, hunting and nest building, but they are nevertheless ultimately inherited from the parents. Mendel's master stroke was to figure out the link between the phenotype and the way genetic material is transferred. That genetic material has been given another name: the 'genotype'.

Genotype

An organism's genotype could also be simply described as its genetic make-up. It is a description of the various genes inherited from its parents. As Gregor Mendel discovered, all organisms get one version of each gene from each of their parents, and so the genotype is made up of these pairs.

A particular genotype does not automatically lead to a related phenotype. In fact, the same phenotype – for example, the tallness of a pea plant – can result from a set of different genotypes (albeit a small set). The mechanisms at play are twofold. Firstly, the different versions of the gene interact and combine with each other in particular ways – described by the ideas of genetic dominance (see page 48) and Mendel's Third Law. Secondly, the environment in which the organism finds itself also has an impact on how it grows and develops, by varying degrees from gene to gene.

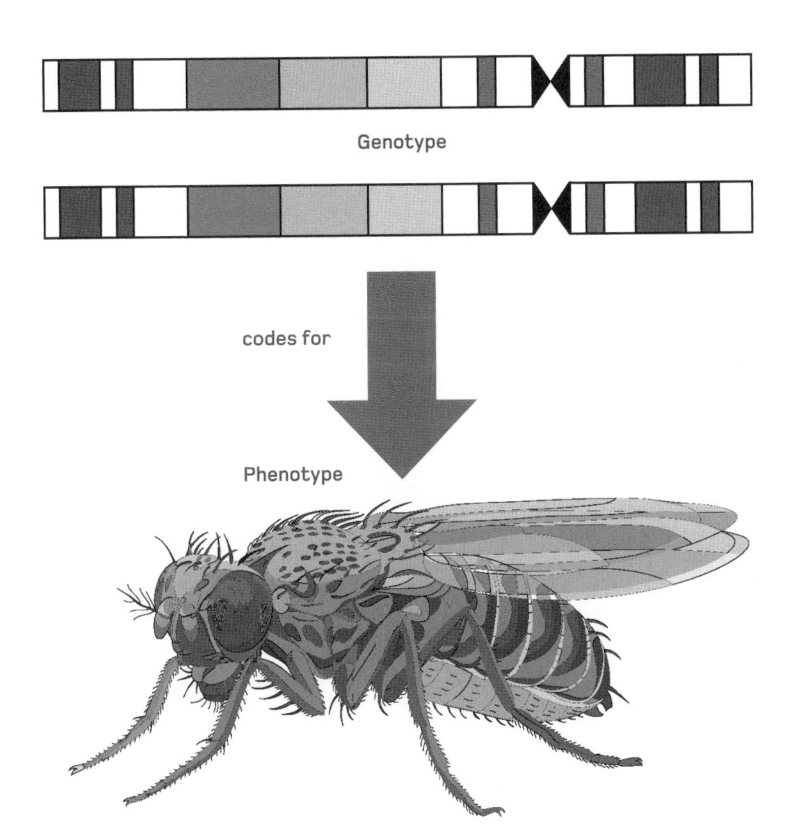
Genotype
codes for
Phenotype

Alleles

This unusual sounding word derives from German and means something like, 'of one another'. It is however a very useful term in genetics: an allele is one of several possible versions of a gene. So using the example of Mendel's pea plants, the gene for plant height has two alleles: tall or short. Another example is eye colour: blue, green, grey, brown and hazel are best described as alleles of the same gene.

A genotype contains two alleles for each gene. If the alleles are identical, then it is described as homozygous – in other words, when it comes to dividing the alleles up to make the sex cells that are used in producing the next generation (pollen, sperm, eggs etc), each cell will definitely contain the same allele. When a genotype contains two differing alleles it is described as heterozygous. As a result, half the sex cells will have one allele, and half the other. Nevertheless, homozygous and heterozygous genotypes can still produce the same phenotype, thanks to an additional complication known as dominance (see page 48).

These guinea pigs all have the same gene that controls hair colour; but each has inherited a different version, or allele, of that gene.

Genome and gene pool

A genome is the total genetic material used by an organism. The concept of our own human genome has become familiar ever since the launch of the Human Genome Project in 1990. That effort to map all of the genetic material used to make a human body was completed in 2003, although the work continues, figuring out how that material is divided up into somewhere between 20,000 and 25,000 genes.

Other organisms with mapped genomes include the *E. coli* baterium, the worm *Caenorhabditis elegans* and the fruit fly. The amount of genetic material – the DNA – in each organism varies, as do the number of genes.

The 'gene pool' is another familiar term but one with a rather different meaning. It refers to the enire set of genes, including their many different alleles, that exists throughout a population of organisms. The gene pool represents the genetic variation wihthin a group of organisms.

As well as studying the genes of an individual, geneticists also seek to understand the behaviours of genes shared by a community, population or entire species.

Hybrids

In general terminology, a 'hybrid' is an organism that is the product of a cross between two distinct breeds. In terms of biology, and genetics in particular, however, a hybrid is an organism that has a *heterogametic* genotype. Put more simply, the organism has inherited two differing alleles, or versions of a gene, from its parents.

Gregor Mendel's success with solving the puzzle of inheritance involved repeated hybridizations. Although he had no idea of how it was happening, Mendel correctly surmised that his crosses of pea plants were creating hybrids – pea plants that had inherited two different versions of a gene.

It was this breakthrough that allowed him to discover that one allele is not always equal to another. Some are dominant over others, and it is this interplay of alleles that shows how a particular genotype leads to a corresponding phenotype.

This żubroń is a hybrid of domestic cattle and European bison.

Dominance

Genetic dominance is the overriding feature that explains the results of Mendel's hybrid experiments (see page 34). Returning to the example of a tall pea plant crossed with a short one, the tall parent has the genotype TT, with T being the tall allele. The short parent's genotype is tt, with t being the short allele.

The next generation of plants all receive a T from one parent and a t from the other. They all have a genotype of Tt. The T allele is dominant over the t, and so all Tt genotypes lead to a tall phenotype. Next, Mendel crossed a Tt plant with itself. This led to four genotypes, all equally likely: TT, Tt, tT and tt. Any genotype with the dominant T allele results in a tall phenotype, while only the tt genotype produces the short phenotype. When explained like that, the 3:1 ratio of tall to short discovered by Mendel makes perfect sense. It also highlights the incredible leap of imagination Mendel made to figure it all out.

Parental generation

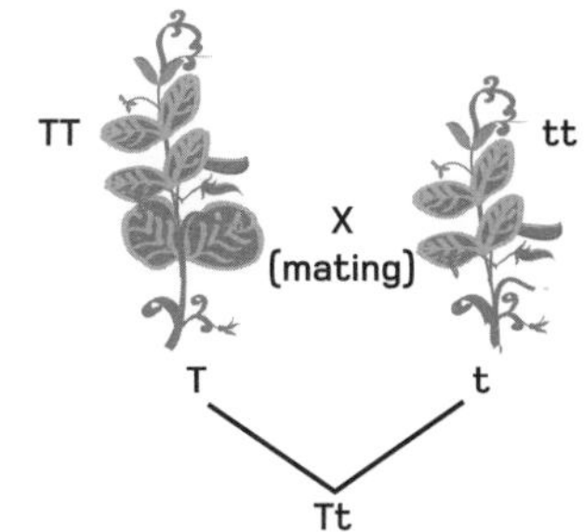

Gametes

All plants in F1 generation have Tt genotype and tall phenotype

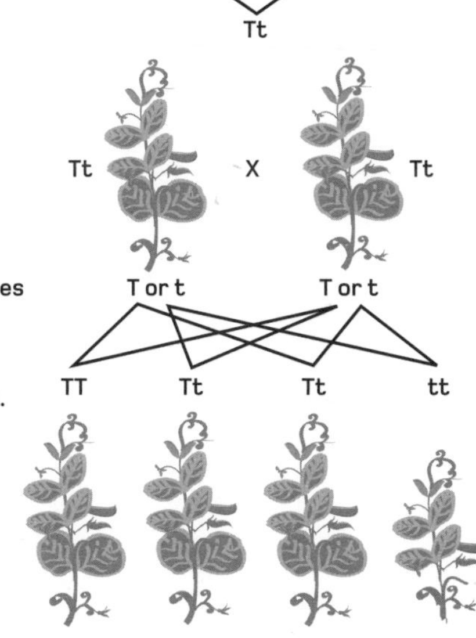

Gametes

Plants in F2 generation have genotypes TT/Tt/tt in ratio 1:2:1. Tall phenotype outnumbers dwarfs by 3 to 1.

Recessive traits

Genetic dominance is not integral to an allele; it is a phenomenon that arises when two alleles meet. One allele may be dominant over the other, in which case this second allele is termed 'recessive' to the dominant one. However, it is possible that the recessive allele in this particular genotype is itself dominant over a third allele. For example, the dark hair allele is dominant over the blonde allele, which in turn is dominant over red hair.

Some traits, such as red hair, are entirely recessive but have little effect on an organism's chance of survival. Other examples, such as albinism (opposite), can be more troublesome. In all cases, recessive phenotypes are infrequent in the population because they require a homozygous, double-recessive genotype. Many genotypes can contain a recessive allele that remains hidden by a dominant partner. Only when two heterozygous parents, or 'carriers', produce offspring is there potential for the recessive phenotype to appear – seemingly out of the blue.

Codominance

The idea of genetic dominance is compelling and simple to understand. However, it is seldom this simple when genotypes are expressed as phenotypes, since there are two other options: codominance and incomplete dominance. These occur when neither of the alleles in a genotype is dominant over the other. The result is that both are expressed in the phenotype in some way, but the difference is a little nuanced.

When a genotype is codominant, both alleles are fully expressed in different parts of the organism. An example would be a red flower and a white flower producing offspring that had flowers covered in red and white blotches. The distinct effects of both alleles are seen. If the genotype is showing incomplete dominance, however, the result is a completely new phenotype that is a blend of the effect of the two alleles. In this situation, a red and a white flower (of a different species to the previous example) would produce offspring with completely pink flowers.

Codominance in cattle

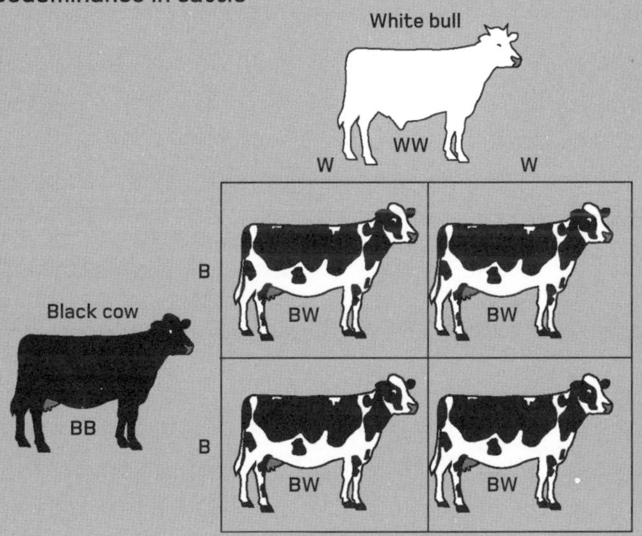

The black and white blotches of dairy cattle are a product of codominance, where two genes are expressed in patches.

The cell

Every living thing is made from at least one cell. The cell is a self-contained packet of life, surrounded by a nanoscopically thin membrane which isolates it from the rest of the its surroundings. Inside the membrane, the cell is filled with a liquid known as the cytoplasm. This is where the cell's metabolism occurs, so it is filled with sugars, proteins and other biochemicals. Also present is some kind of genetic material in the form of DNA. The field of cell biology investigates every aspect of this tiny world, and in so doing has revealed much that all cells have in common, and much that is specific to different types of organism.

The word 'cell' was coined by the great English scientist Robert Hooke, who was first to discover these tiny structures. In 1665, Hooke peered at many life forms through what was then a cutting-edge microscope. When he looked at a slice of cork he found it divided into tiny compartments (opposite). He likened these to the living quarters of a monk, known then as a cell.

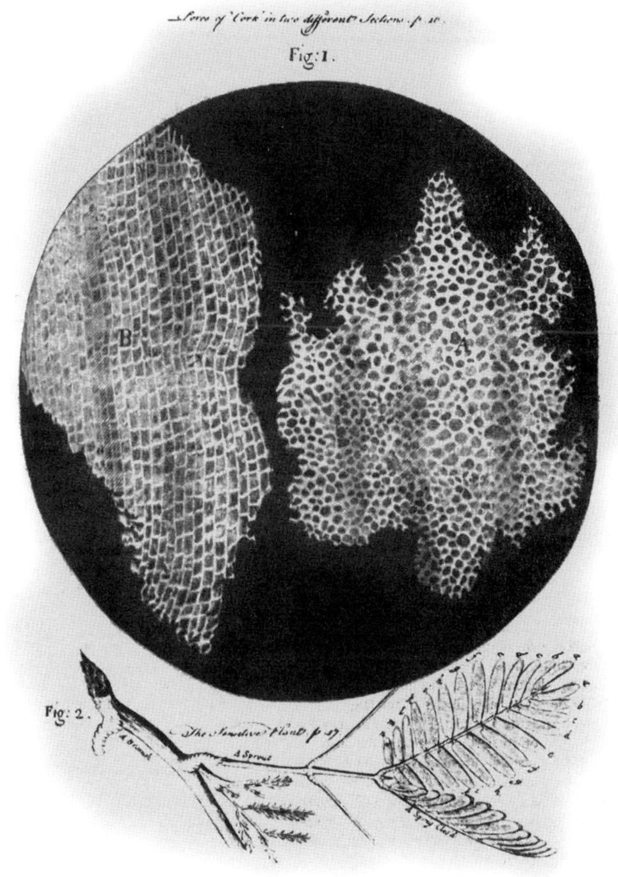
Fig: 1.
B
A
Fig: 2.
A Sprout

Cell membranes

The outer covering of a cell is called its plasma membrane. There are similar membranes inside many cells, and they all share a structure based on the properties of chemicals called lipids. As a class of chemicals, lipids are perhaps better understood as animal fats and vegetable oils.

Each lipid molecule has a hydrophobic side that repels water, and a hydrophilic side that dissolves in water. A cell membrane is made of a double layer of these molecules – the hydrophilic parts face outwards to form the inner and outer surface of the membrane, while sandwiched between them, the hydrophobic parts mingle with each other. This creates a barrier that is surprisingly strong when constructed at the tiny dimensions of a cell. Water can move freely across the membrane, but most larger molecules must be actively transported into and out of the cell. For this purpose, the cell membrane is studded with pores and pumps made by complex protein molecules (see page 112).

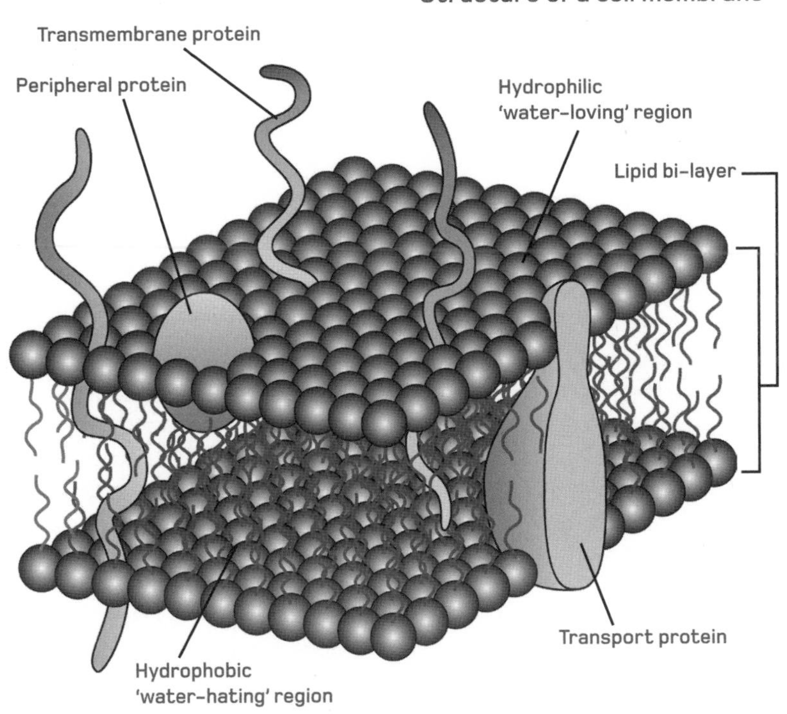
Structure of a cell membrane
Transmembrane protein
Peripheral protein
Hydrophilic
'water-loving' region
Lipid bi-layer
Transport protein
Hydrophobic
'water-hating' region

Prokaryotes

The smallest and simplest cells belong to bacteria and their cousins the archaea. These organisms are the prokaryotes, and their cells are very different from those of other organisms, such as plants, fungi and animals, which are known as the eukaryotes.

Most prokaryotic cells are somewhere between 1μm (1 micrometre – a millionth of a metre) and 5μm long. They are mostly limited to this size range by their plasma membranes (some bacteria have two membranes around the cell), which are less fluid and flexible than in other cell types. Some cells have a long twisted tail-like extension called a flagellum, which spins like a corkscrew to propel the cell. There are smaller hair-like extensions called pilli, which cling to surfaces. Internally, the cell appears quite spartan: a tangle of DNA molecules floats freely in the cytoplasm, and the only other structures visible are ribosomes – sites where the genetic code is read and processed (see page 104).

Prokaryotic cell

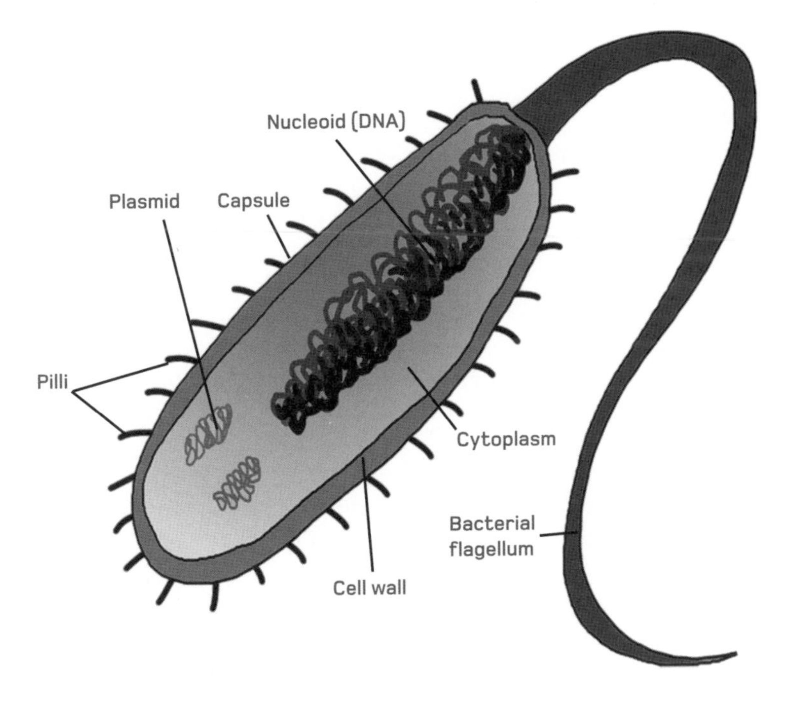

Animal cells

The cells in your body share the same structures as those in all animals and many single-celled eukaryotes, such as amoebae. The animal cell is around 20 times the length of a prokaryotic cell (and with a considerably larger volume) – mostly thanks to a cell membrane that is strengthened by the inclusion of cholesterols. An animal cell may use a flagellum (plural: flagella) to move, and it may also be equipped with cilia – shorter extensions that are similarly mobile. These are used for locomotion as well as for drawing a nutrient-rich current over the cell.

The larger size of eukaryotic cells means they cannot rely on passive diffusion alone to distribute material in the cytoplasm. Instead, there is a network of microtubules to transport useful molecules around. The cell is also filled with an array of structures called organelles, each controlling an aspect of its metabolism. The final distinction from bacterial cells is that the DNA is held in a nucleus.

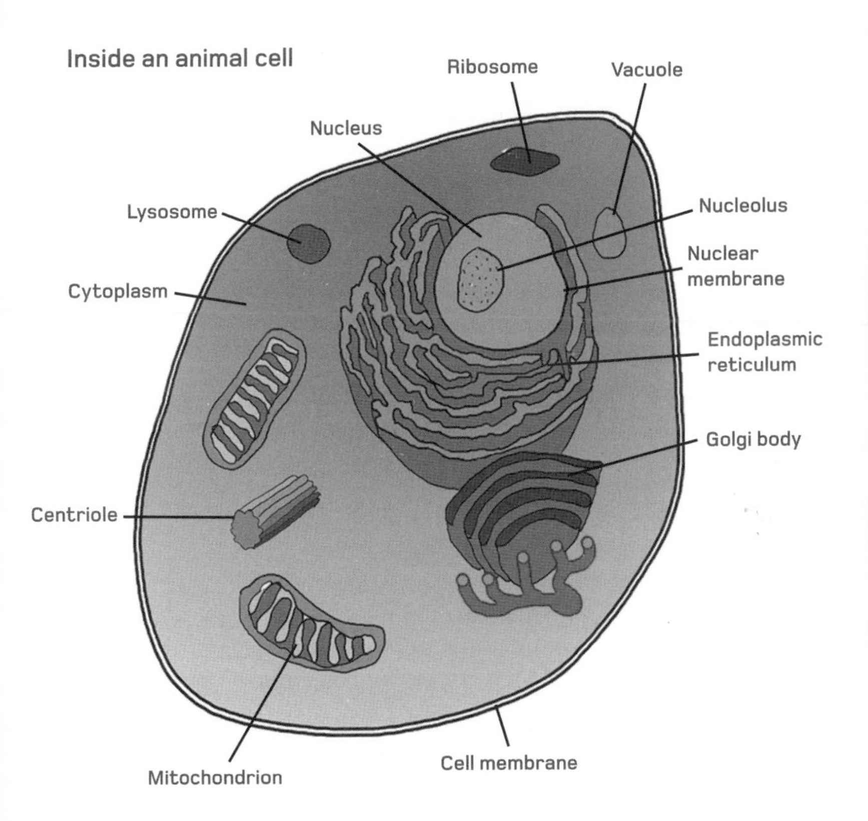
Inside an animal cell
Ribosome
Vacuole
Nucleus
Lysosome
Nucleolus
Nuclear membrane
Cytoplasm
Endoplasmic reticulum
Golgi body
Centriole
Cell membrane
Mitochondrion

Plant and fungal cells

As another example of an eukaryote, a plant cell shares many features with the animal cell. There is a nucleus containing the genetic material, and a similar collection of organelles working away inside the cell membrane. One major difference is the inclusion of chloroplasts – the structures in which photosynthesis takes place.

The other big difference is that a plant cell has a cell wall that surrounds the cell membrane. While the flexible membrane is able to change volume as water flows in and out of the cell, the wall is stiff and unchanging. A plant cell wall is made from fibres of cellulose, a polymer of sugar molecules linked into a chain.

Fungal cells could be seen as a halfway house between plants and animals (although they are closer relatives of animals). The cells do not photosynthesize and so lack chloroplasts, but they do have a cell wall. Instead of cellulose, this wall is made from chitin, a polymer also found in seashells and insect bodies.

Inside a plant cell

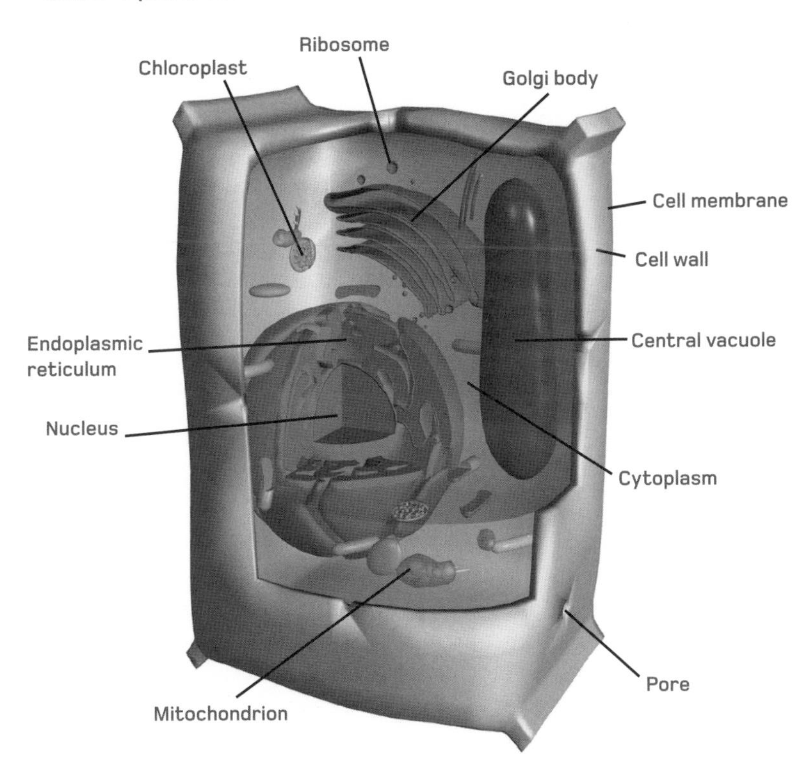

Multicellularity

The most familiar organisms are obviously made up of many cells – often numbering in the trillions – working together. This multicellular mode of life is in contrast to the unicellular mode, where bacteria and a range of eukaryotic organisms can survive as single cells.

However, the distinction is not completely clear-cut. Many unicellular organisms form colonies – a throat infection is a colony of bacteria living in your larynx. There are many cases documented where colonies of unicellular organisms exhibit a division of labour with certain cells specializing in tasks that support the wider colony.

A multicellular body takes this further, with genetically identical cells specializing in several ways to ensure the survival of the whole. One of the simplest examples of this is the sponge, an animal that uses nine cell types to provide its bodily structure, feed, defend and reproduce.

Sponges and other simple sea life represent the first stage in animal evolution, where collections of different cells work together to make a body.

Nucleus and organelles

The invention of the electron microscope in the 1930s revealed that the interior of cells was more complex than anyone had previously suspected. Until then the only thing clearly visible inside eukaryotic cells was the nucleus. However, the greater resolution offered by the new technology showed this in much more detail, and also revealed several distinct structures, named organelles. The nucleus was shown to have not one but two membranes around it, both riddled with pores that allow genetic material to be hauled in and out. The store of genetic material is massed in a central region called the nucleolus.

The other organelles include the endoplasmic reticulum, which is a transport network of tubes. The Golgi apparatus prepares material for release from the cell, while the lysosome is used to collect and destroy unwanted material. Plant cells also have chloroplasts, while all eukaryotes have mitochondria, which act as the power plants of cells.

The Golgi apparatus

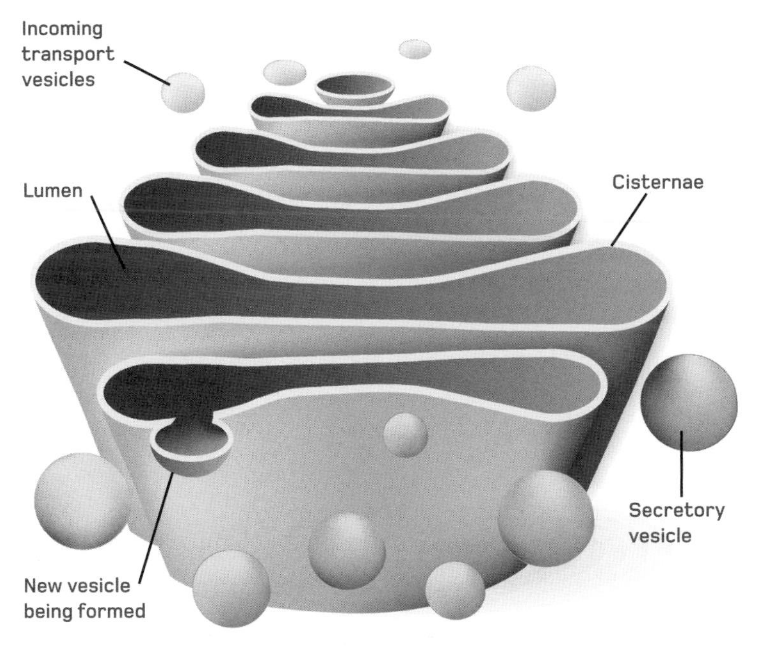

The Golgi apparatus parcels up chemicals into vesicles which then merge with the cell membrane, releasing the chemical outside the cell

Mitochondria

The mitochondria (singular: mitochondrion) are the places in a cell where respiration occurs, releasing energy from glucose and other fuels in manageable amounts, for use in metabolism. Every eukaryotic cell has some mitochondria, and those that use a lot of energy, such as muscle cells, are packed with hundreds.

A mitochondrion has an outer membrane similar to the plasma membrane of the overall cell, with another inside that is folded in on itself. creating many narrow dead-ended channels called cristae. Respiration takes place on the membrane walls of the cristae. The reactions that occur here gradually oxidize a glucose molecule, and the energy released at each step is captured for future use by a chemical called ADP (adenosine diphosphate). An input of energy allows another phosphate to be added to this in order to create ATP (adenosine triphosphate). The ATP molecules then head out into the cell, and can release their stored energy whenever needed by reverting to ADP.

Inside a mitochondrion

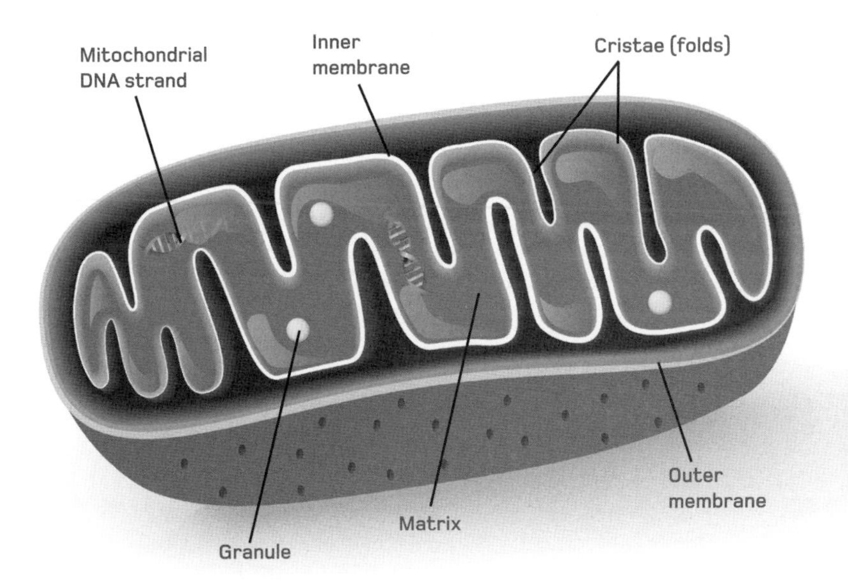

Chloroplasts

The green-coloured organelles called chloroplasts are the sites of photosynthesis inside plant cells. Those parts of a plant that do not photosynthesize, such as the roots, lack chloroplasts, as do photosynthetic prokaryotes (although the process is largely similar there, it takes place in the cytoplasm along with all the other metabolism).

A chloroplast has its own outer membrane, containing further membranous discs known as thylakoids. Chlorophyll molecules are bonded to individual thylakoids, which are stacked up into structures called grana. When light shines onto them, the chlorophylls convert some of its energy into a supply of ATP molecules (adenosine triphosphate – see page 68). This is the first stage of photosynthesis and is named the 'light reaction'. The ATPs then fuel the second stage, or 'dark reaction'. This takes place in the stroma, the space between the grana, and involves the combination of carbon dioxide and water to produce glucose sugars.

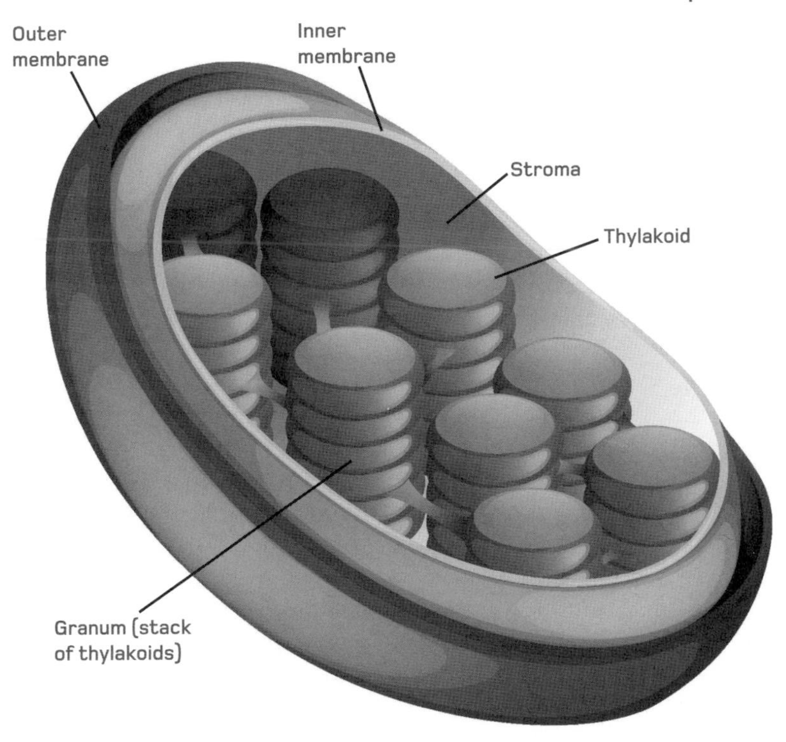
Inside a chloroplast
Outer membrane
Inner membrane
Stroma
Thylakoid
Granum (stack of thylakoids)

Cell theory

When the first single-celled organisms (initially called 'animalcules') were discovered by the pioneers of microscopy, biologists assumed that these organisms developed spontaneously from the decaying remains of other life forms. It was well understood that multicellular life arose through biogenesis – one life creating another – but unicellular life was thought to be abiogenic, arising from non-living material.

In 1838, however, German physiologist Theodor Schwann (with contributions from others) used the growing evidence against abiogenesis to propose what became known as 'cell theory'. Schwann's theory had three parts: first, all organisms are composed of one or more cells; second, the cell is the simplest form of life; and finally, all new cells arise from older ones. These three simple rules became the foundations of modern biology and were crucial in figuring out the genetic process.

Theodor Schwann's inspiration for cell theory began when he noticed the shared similarities between nerve, muscle and plant cells.

Chromosomes

In accordance with cell theory, every new cell arises from an older cell splitting in two. This cell division allocates the contents of the original cell more or less equally, and thus requires the nucleus to divide in two as well. Early studies of this process revealed that the nucleus was filled with a coloured material, which was named chromatin. In 1888, Heinrich Wilhelm Gottfried von Waldeyer-Hartz saw that prior to division, the diffuse chromatin formed into fibres that were then divided between the two new cells. He named these 'chromosomes'.

Today, we know that chromosomes are the scaffold structures holding the DNA molecules that make up the genome. Human cells have 46 chromosomes, but numbers vary wildly from species to species. Most of the time, individual chromosomes are too narrow to see, existing as a nebulous mass of chromatin, with the DNA coiled around spindle-like proteins called histones. Only during cell divisions will the coils thicken into structures visible through microscopes.

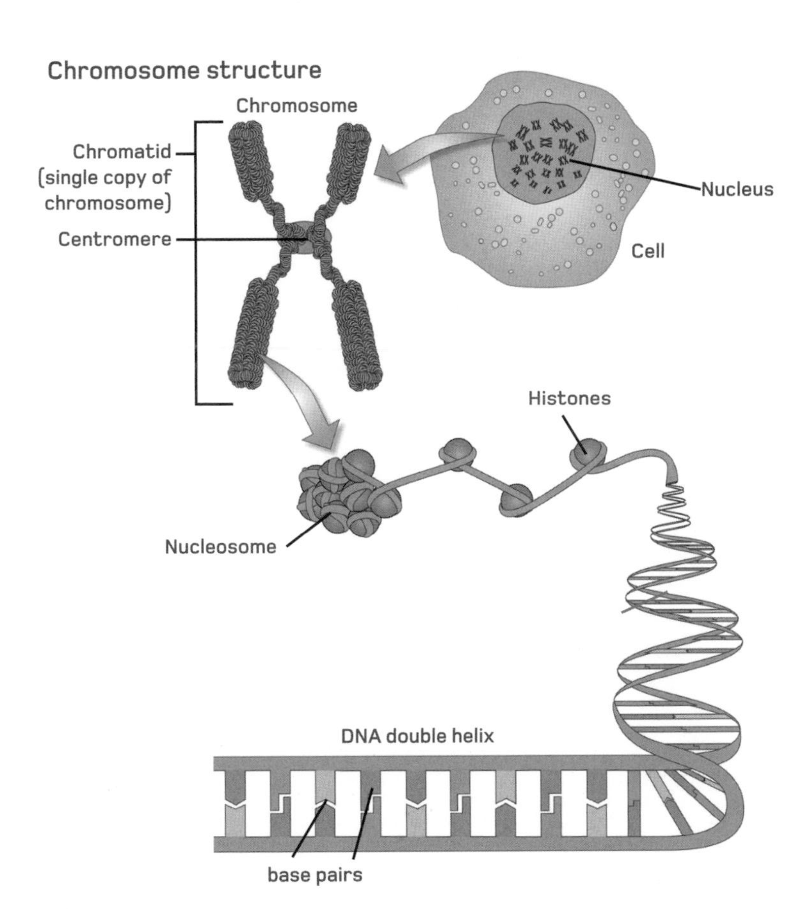

Chromosome structure
Chromosome
Chromatid
(single copy of
chromosome)
Centromere
Nucleus
Cell
Histones
Nucleosome
DNA double helix
base pairs

Cell division: Mitosis

The main process of cell division – the one used to grow the bodies of individuals – is called mitosis. This name, derived from the Greek word for weaving, was coined because mitosis involves a network of threadlike microtubules that are seen to pull the chromosomes into two groups and heave them to either ends of the cell prior to the division.

Mitosis occurs over several complex stages, but to summarize, the first step is to duplicate each chromosome strand. This results in the X-shaped structure commonly associated with chromosomes, which is in fact a pair of chromatids – identical copies of the chromosome connected together. Once the chromosomes have duplicated, the nuclear membranes dissolve, allowing the chromosomes to line up across the cell. The microtubules then pull the chromatids apart, temporarily doubling the number of chromosomes in the cell. Finally, a new plasma membrane develops down the middle of the cell, allowing it to cleave in two.

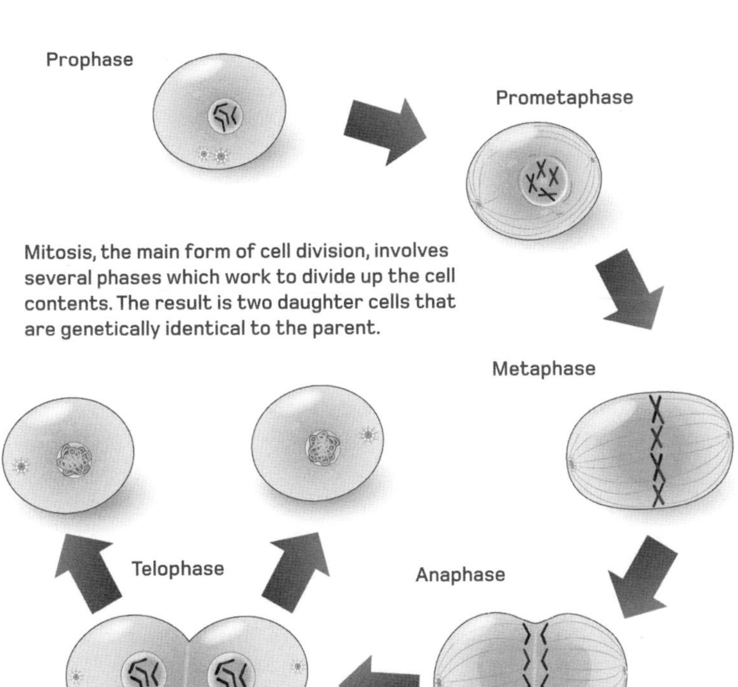

Mitosis, the main form of cell division, involves several phases which work to divide up the cell contents. The result is two daughter cells that are genetically identical to the parent.

DNA

The most famous three letter acronym in science is short for 'deoxyribonucleic acid'. DNA is the most important chemical in genetics, because it is within the intricate structure of its long molecules that the code for all our genes is held. Every person has 2 metres (6 ft 6 in) of DNA in each of their cells. – added together, the total DNA in a human body would stretch to the Sun and back 66 times. The vast majority of a cell's DNA is housed in the nucleus, with a tiny amount in the mitochondria.

DNA was first isolated in 1869 when Swiss physician Friedrich Miescher (opposite) analysed pus collected from bandages covering infected wounds. It was found to contain the sugar ribose, phosphates and nitrogen-containing organic acids. It was soon firmly associated with the contents of the cell nucleus, and its acids were named nucleic acids accordingly. In 1928, DNA was found to reside in chromosomes, confirming that it was indeed the long-sought genetic material that carried genes between generations. The question was: how did it do it?

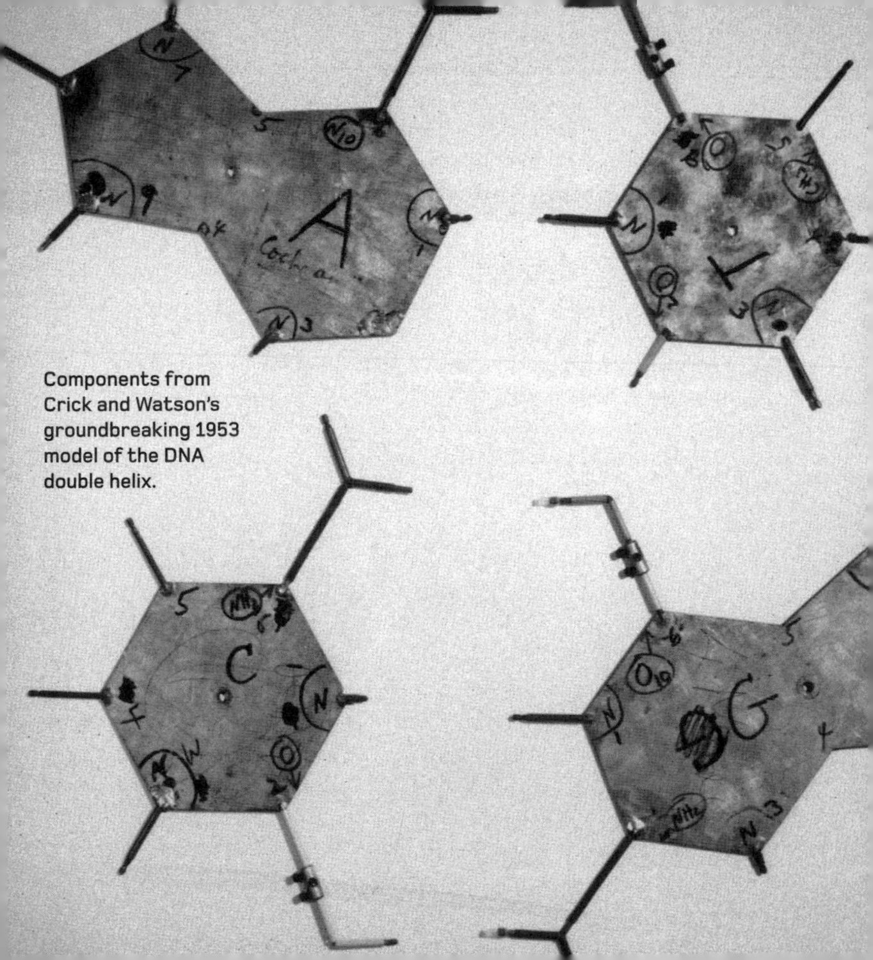

Components from Crick and Watson's groundbreaking 1953 model of the DNA double helix.

X-ray crystallography

To understand how DNA carries genes, it was first necessary to figure out its structure. DNA is a polymer, meaning it is made up of many smaller units chained together. Those units were recognized, but it was not understood how they linked. Individual DNA molecules are too small to see directly, so it was not until the 1950s that the pioneering technique of X-ray crystallography was able to solve the puzzle.

X-ray crystallography made use of the phenomenon of diffraction, a process common to waves of all types. When a wave passes through a gap that is narrower than its wavelength, it propagates from the gap in all directions as if it was starting again from a point source. X-rays have a small enough wavelength to diffract as they shine through the spaces in a molecule. The diffracted waves that emerge on the other side interfere with each other to create a pattern of light and dark bands that can be interpreted to pinpoint the relative positions of gaps in the molecule, and reveal its shape.

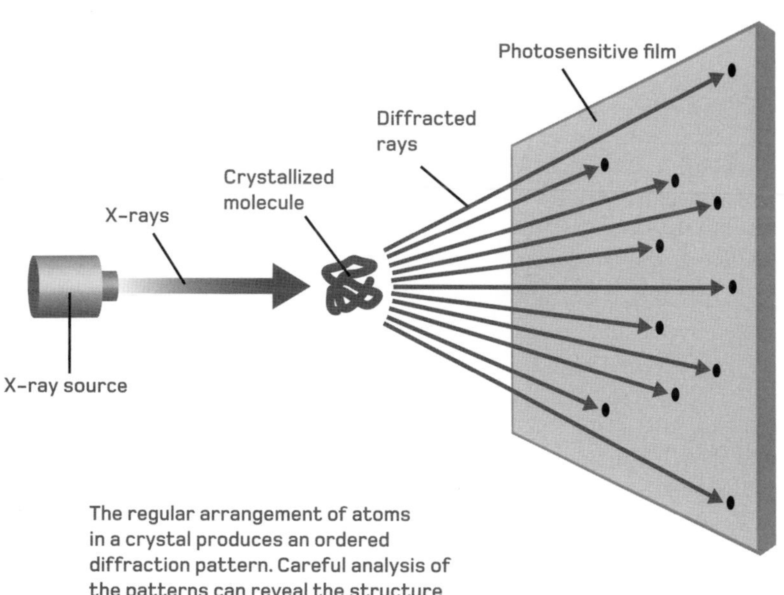

The regular arrangement of atoms in a crystal produces an ordered diffraction pattern. Careful analysis of the patterns can reveal the structure of organic molecules, such as DNA.

The double helix

The first X-ray crystallography results for DNA were interpreted by the American Nobel-Prize winning chemist Linus Pauling in 1951. He proposed that DNA molecules had a spiral structure he named an alpha helix. This proved to be not quite right, however, and in 1953 two researchers working in Cambridge – James Watson and Francis Crick (see page 84) – showed that the molecule was actually a *double* helix, composed of two strands of DNA that linked together.

The DNA double helix can be understood as a twisted ladder. Its two 'uprights' are made from chains of ribose sugars bonded together by phosphates, while the 'rungs' are pairs of nucleic acids, now known simply as bases, that are strung between the chains of sugar. This unique structure allows DNA to be separated into two distinct strands by splitting the base pairs in two. There are four base units used in DNA, and it is their order through the double helix that carries the coded information of every gene.

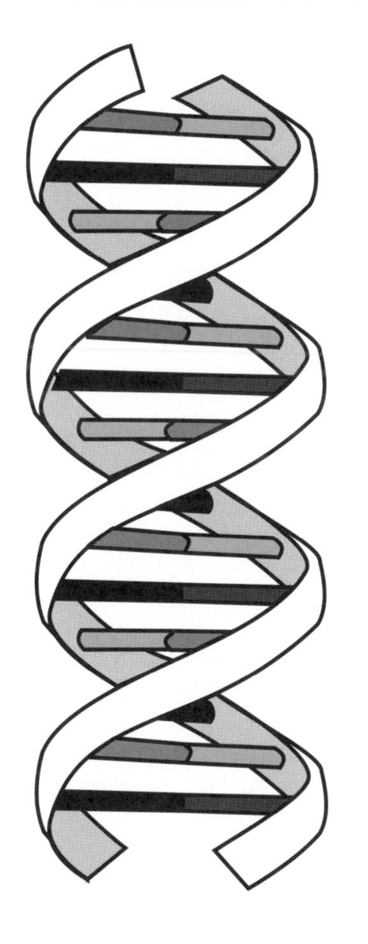

Adenine

Thymine

Cytosine

Guanine

Sugar-phosphate
backbone

Crick and Watson

Francis Crick and James Watson won the Nobel Prize for Medicine in 1962 for their discovery, a decade earlier, of the double helix structure of DNA, along with their work in showing how it could be used to store genetic information.

Englishman Crick (opposite, right) had worked as a physicist and weapons designer during the Second World War, before turning to biology, where he developed mathematical theories for use in interpreting X-ray diffraction. Watson (left), an American molecular biologist, joined Crick at Cambridge's Cavendish Laboratory, where they collaborated to deduce the now famous structure of DNA. They did not perform their own X-ray diffraction experiments, but were instead supplied with data by Maurice Wilkins, the supervisor of a lab at King's College, London. Wilkins shared the 1962 prize with Crick and Watson, even though he had not collected the images himself – the crucial evidence was gathered by his late colleague Rosalind Franklin. She was never consulted, and controversy over her role has raged ever since.

Photo 51

One particular X-ray crystallography photograph was pivotal to allowing Crick and Watson to solve the mystery of DNA's molecular structure. The pattern of light and dark regions in the famous 'Photo 51' provided the evidence they needed to confirm their idea that a double helix was indeed the true form of the DNA found in chromosomes.

Photo 51 is often attributed to Rosalind Franklin, but it was actually produced by Raymond Gosling, a PhD student working under her supervision at King's College, London. Franklin had filed the photo for later study, and it was then given to Maurice Wilkins by Gosling several months later. Wilkins showed it to James Watson, who immediately saw its potential – in the late 1940s, chemist Linus Pauling and his team at the California Institute of Technology had used the X-shaped diffraction patterns produced by many proteins to prove that they contained helical shapes, and the strong X-pattern in Photo 51 suggested that DNA too was a helix.

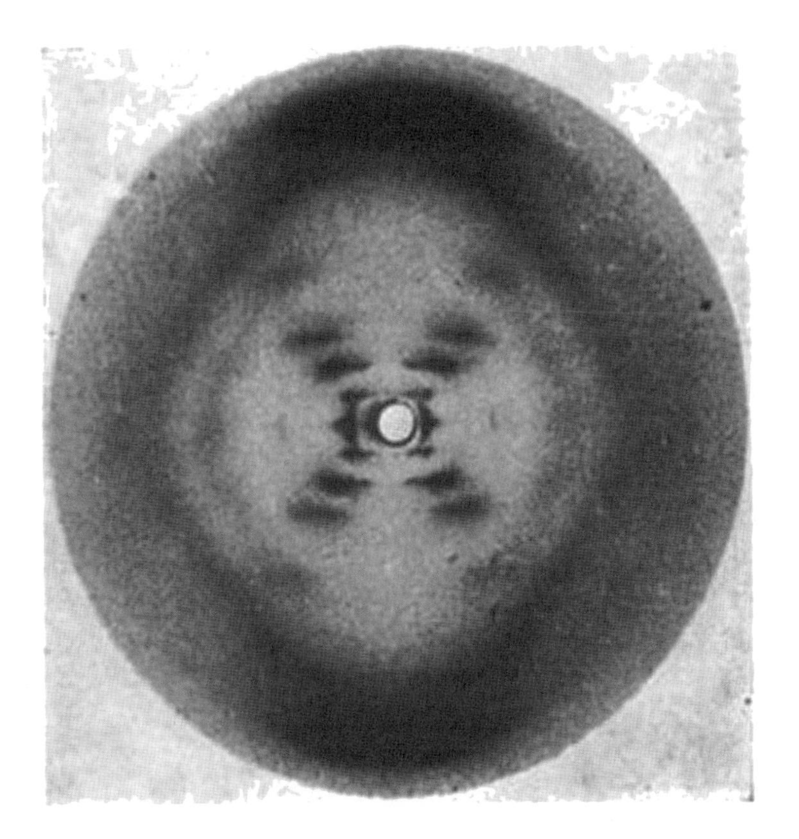

Rosalind Franklin

By the time the Nobel committee chose to award Crick, Watson, and Franklin's erstwhile colleague Wilkins a prize in 1962, Rosalind Franklin was already dead. She had passed away in 1958 at the age of just 37, killed by ovarian cancer that may have been brought on by her repeated exposures to dangerous levels of X-rays. The Nobel rules dictate that a prize is never awarded posthumously, so we will never know whether, if Franklin had lived, she would have joined the three men as an equal winner.

In later years, Franklin's contributions to DNA research have been more widely acknowledged. Her major breakthrough was not Photo 51 *per se*. Instead, she had discovered that the DNA found in a cell's nucleus is in a highly hydrated form, which she called B-DNA. (A-DNA is a less hydrated form that is uncommon in biology.) Crick and Watson's eventual success came from modelling the structure of this B form of DNA, a right-handed double helix.

Base pairs

A DNA double helix contains millions of 'base pairs', each consisting of two nucleic acids bonded together that run across the axis of the helix. Crick and Watson's discovery of the double helix showed that the base pairings formed according to a set of simple rules.

There are four bases in DNA: adenine, cytosine, guanine and thymine. Each of them consists of ringed organic compounds containing nitrogen atoms. Adenine and guanine share a similar structure, having two rings, while cytosine and thymine have just one apiece.

For a base pair to fit across a 'rung' of the helix molecule, however, there is only room for three rings. Adenine always pairs with thymine, while guanine pairs with cytosine. There is no restriction on the direction of the pairing. Any of the four bases can appear on one side of the helix, but its partner will always be on the other side.

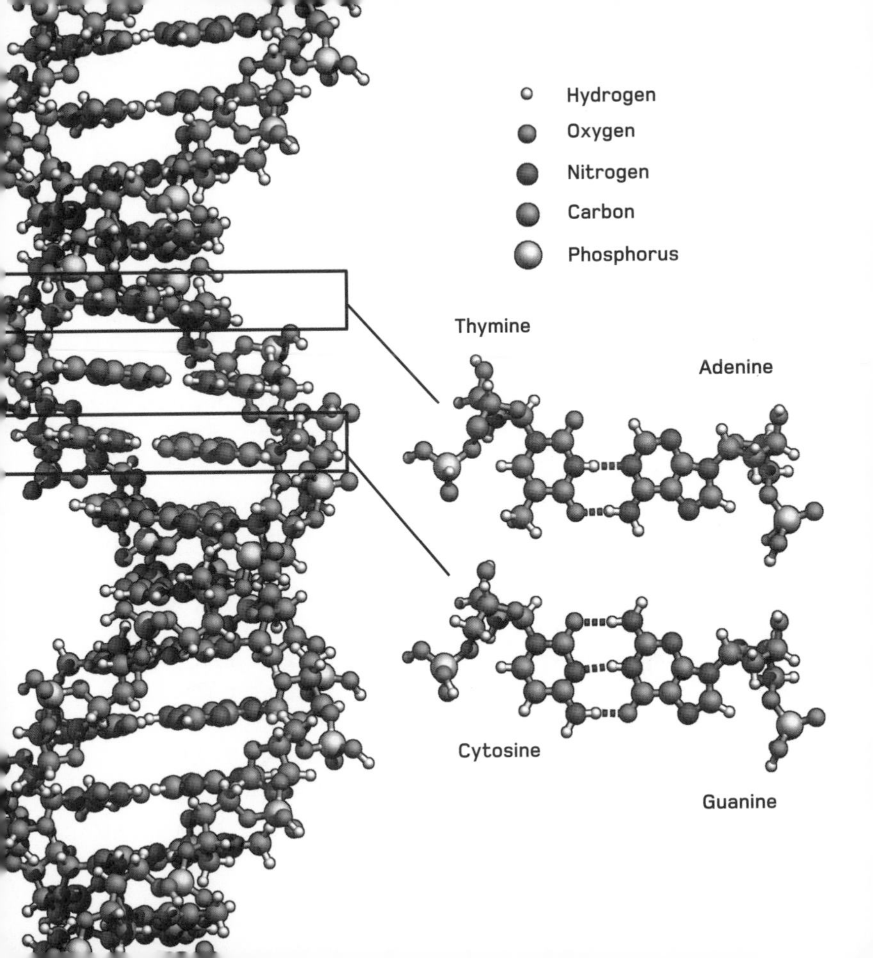
Hydrogen
Oxygen
Nitrogen
Carbon
Phosphorus
Thymine
Adenine
Cytosine
Guanine

Genetic code

The four DNA bases – adenine, cytosine, guanine and thymine – create a four-character code, generally abbreviated to the initial of each base: ACGT. The Human Genome Project has found that the DNA in a human cell contains a code of more than 3 billion of these characters – printed out, they fill 130 printed volumes of solid text. By reading one letter a second, it would take 95 years to read the whole thing – and much of its meaning is still a mystery.

Because each DNA molecule is a double helix, it is built from two strands of DNA entwined together. Each strand carries a code using the ACGT alphabet, running to millions of characters long. What's more, the code on one strand is always mirrored on the opposite strand that carries the partner bases. However, only one of these strands carries the 'live' genetic code used by cellular processes – this is known as the 'sense' strand, while the opposite one is the 'antisense' strand.

CCCADTCCAGGATCCADTCCAGAT
CGGATCCADTCCAGGATCCADTCC
TCCATCGATCCADTCCAGGATCCA
CADTCCATCGGATCCADTCCAGGA
GGATCCADTCCATCGGATCCADTC
DTCCAGGATCCADTCCATCGGATC
ATCCADTCCAGGATCCADTCCATC
CGGGATCCADTCCAGGATCCADT
DTCCACGGGATCCADTCCAGGAT
CCADTCCACGGGATCCADTCAGG
AGGATCCADTCCACGGGATCCAD
ADTCCAGGATCCADTCCACGGGA
ATCCADTCCAGGATCCADTCCACG

Self-replication

During every cell division, all the DNA in the nucleus must be copied or replicated. The replication process makes use of the fact that one side of the DNA double helix can act as a template for its partner. Replication can take place through purely chemical means in a process called autocatalysis, but inside the cell it is heavily managed by enzymes to ensure that a high-fidelity copy is produced. The double helix is unzipped into two strands. Individual bases are paired up with their partners on each exposed strand, and a new ribose backbone is constructed to keep them in place. The result is two identical double helices. 'Proof-reading' enzymes check the pairings to ensure that there are no errors.

At the level of the chromosome, this results in two structures, known as chromatids, that carry identical copies of DNA. The chromatids are connected at a point called the centromere. When the chromatids separate during cell division, they become individual chromosomes in their own right.

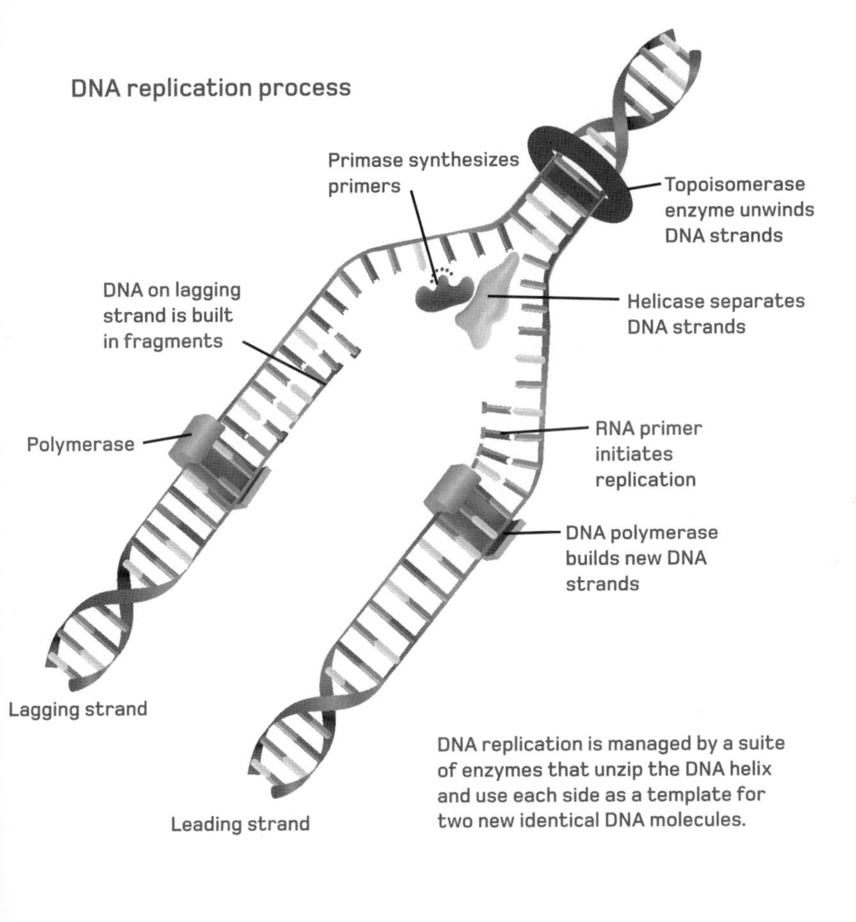

DNA replication process

DNA replication is managed by a suite of enzymes that unzip the DNA helix and use each side as a template for two new identical DNA molecules.

Junk DNA

During the replication process, all the DNA in a cell is involved. However, when DNA code needs to be read by the cell, only certain portions of the strands are used – the rest is simply ignored. The sections of a double helix that contain a meaningful code are known as exons, while the unused portions are called introns.

Introns of varying lengths are muddled among the exons along the DNA. The introns are often termed 'junk DNA', suggesting that they have no use – they are simply there because the cell faithfully copies it along with all the other bits of DNA. However, the preferred scientific name for it is non-coding DNA, since its actual utility is still much debated – perhaps it has some as yet unknown purpose? According to the latest estimates, more than 90 percent of human DNA takes the form of non-coding introns. This figure is much higher than in simpler organisms such as bacteria, where barely 10 per cent is intronic, so it may be that introns arise naturally as DNA grows more complex.

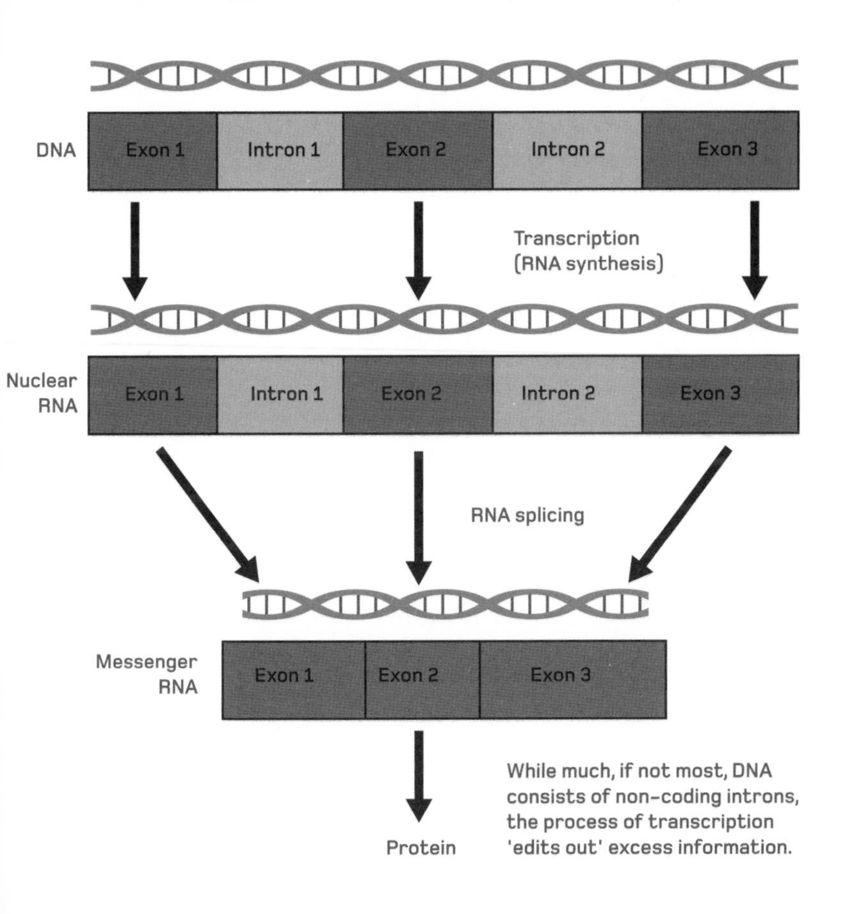

While much, if not most, DNA consists of non-coding introns, the process of transcription 'edits out' excess information.

The Central Dogma

The discovery of the double helix was merely the first step in solving the mystery of genetic inheritance. Over the ensuing decades, a picture was gradually pieced together to show how the information coded in DNA is deciphered and enacted by the cell. This awe-inspiring mechanism has become known as the 'central dogma' of molecular biology.

The central dogma equates a particular portion of DNA, known as a cistron, with a particular protein used by the living cell. Each cistron contains the information to build one single type of protein. This means that 'cistron' and 'gene' are two words for the same thing – a single unit of inherited information.

Secondly, the dogma shows that the information on the cistron only travels in one direction: the DNA code can be used to create a protein, but the structure of the protein cannot be translated back into a DNA code.

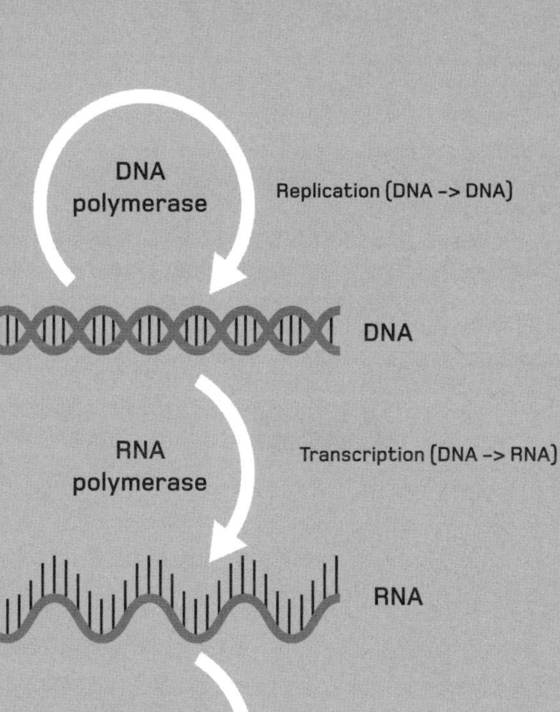
DNA
polymerase
Replication (DNA -> DNA)
DNA
RNA
polymerase
Transcription (DNA -> RNA)
RNA
Ribosome
Translation (RNA -> Protein)
PROTEIN

RNA

While DNA is the repository of the information that forms a cell's genes, the cell actually uses a related chemical called RNA to read it. RNA is short for 'ribose nucleic acid'. Its molecules are broadly similar to DNA in their fine structure, save for the different way in which the sugar backbone is constructed from ribose sugar molecules linked by phosphate groups. The presence of a hydroxyl (OH) group attached to the ribose ring affects the structure and makes it impossible for RNA strands to form into long double helices. Instead, they are more often found as robust clusters of short helices or as single strands. RNA uses three of the same bases as DNA: adenine (A), cytosine (C), guanine (G). But in place of thymine (T), RNA uses a base called uracil (U). Therefore, the genetic alphabet of DNA – ACGT – translates into ACGU when RNA is involved.Thanks to its robust nature compared to the more fragile DNA molecule, RNA is the workhorse of the central dogma process that copies, transports and reads genetic information within every cell.

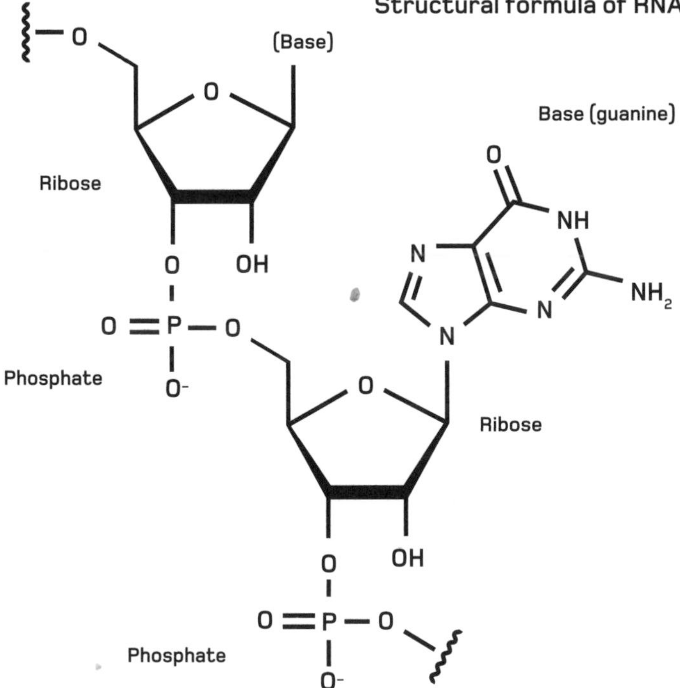

Structural formula of RNA
(Base)
Base (guanine)
Ribose
Phosphate
Ribose
Phosphate
O
OH
O
P
O
O⁻
NH
N
NH2
N
N
OH

Transcription

While genes are stored in the nucleus, the proteins they code for are assembled outside, in the cell's cytoplasm – and specifically at tiny organelles called ribosomes. As a result, genes need to be copied and transported from the nuclear DNA to these external sites. The copying process is called transcription, and unlike the complete replication of the DNA helix, it results in a single copied strand – of RNA, rather than DNA. The aim of transcription is to make a copy of the 'sense' strand – the side of the DNA helix that carries the protein-building instructions. Therefore, the helix is unzipped and the opposite 'antisense' strand is used as a template for the resulting piece of RNA.

This copy contains all the introns in the gene as well as the exons, and the introns are snipped out to create an edited copy. This finished strand, manufactured in the nucleus, is called messenger RNA (mRNA). It is then hauled through a pore in the nuclear membrane and begins its journey to the ribosome.

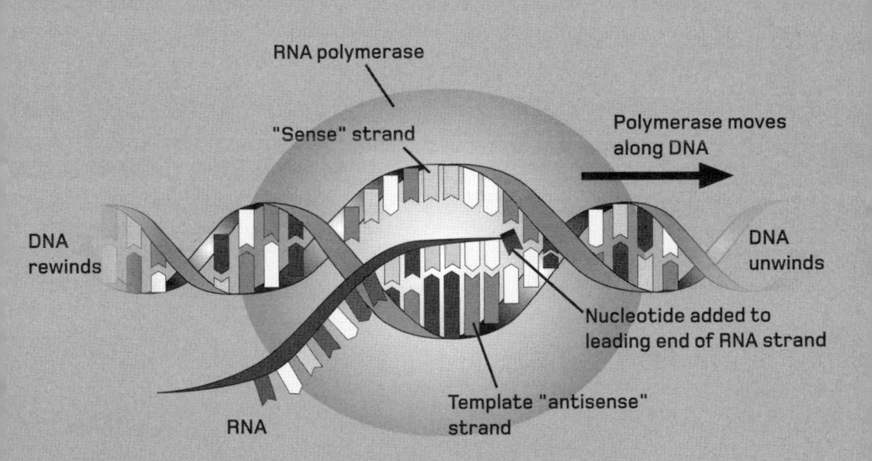
RNA polymerase
"Sense" strand
Polymerase moves along DNA
DNA rewinds
DNA unwinds
Nucleotide added to leading end of RNA strand
Template "antisense" strand
RNA

Ribosomes

The minute bodies known as ribosomes are often lumped in with other cell machinery under the catch-all term of organelles, but they are in fact distinctly different, since they are found in both eukaryotic and prokaryotic cells (and the latter by definition do not have any of the large internal structures classed as organelles). The widespread distribution of ribosomes is a pointer to their significance in cell biology, and their primordial nature – they are generally presumed to have evolved before the major evolutionary separation of eukaryotic and prokaryotic organisms.

A ribosome is a site where the information coded on a strand of mRNA (messenger RNA) is read and used to assemble a protein. The ribosome itself is largely made from another form of RNA, known appropriately as ribosome RNA or rRNA. The rRNA forms two subunits, one large and one small. Every strand of mRNA is threaded between these two units during a reading process known as translation.

Structure of the ribosome

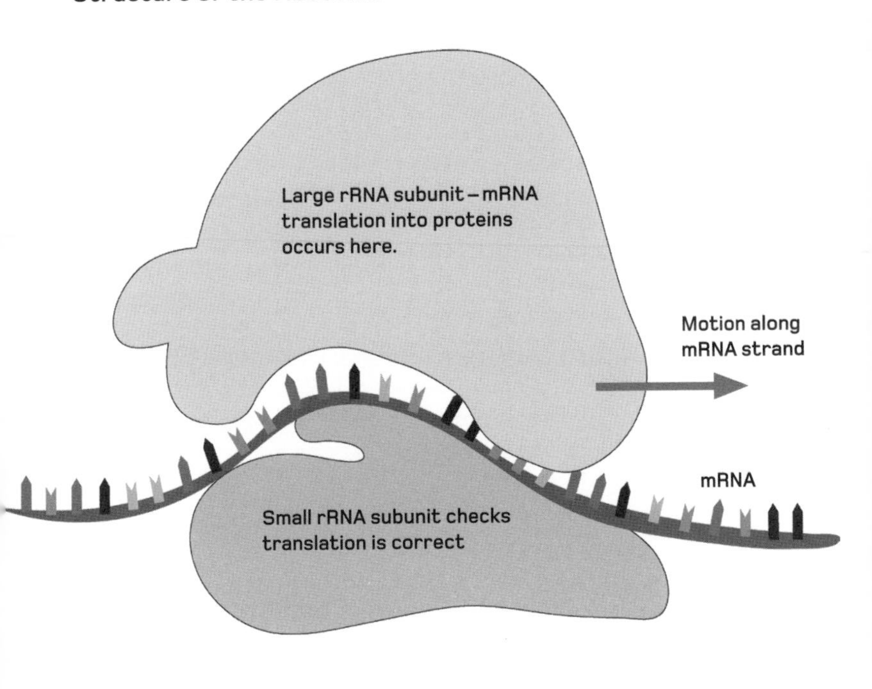

Translation

Protein synthesis involves two key phases – the copying of DNA information onto messenger RNA in the cell nucleus (transcription), and the formation of a protein using that information, known as translation. Ribsomes expose mRNA to a reading system composed of yet more RNA structures – this time made of transfer RNA. This tRNA is a loop of RNA with three bases exposed at one end. These bases correspond to three partner bases on the mRNA held in the ribosome. Each of the mRNA's three-letter codes is called a codon, while the corresponding three letters on a tRNA is its anticodon.

MRNA passes through the ribosome one codon at a time, allowing a tRNA to match up with each one. Attached at the other end of the tRNA is an amino acid, the building block of a protein molecule. Each codon and anticodon pairing corresponds to a specific amino acid, in a specific position among the hundreds that are required to build a particular protein. The protein is assembled by reading all the gene's codons in the right order.

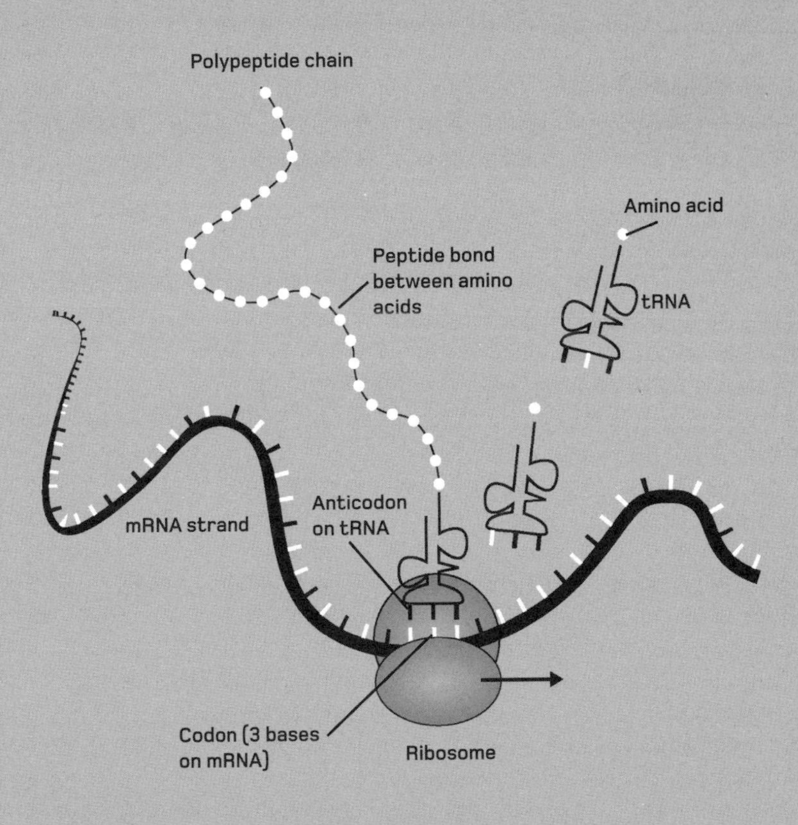
Polypeptide chain
Amino acid
Peptide bond between amino acids
tRNA
Anticodon on tRNA
mRNA strand
Codon (3 bases on mRNA)
Ribosome

Codons

The codon is the interface between the genetic world and the metabolic one. A codon is a simple three-letter string of bases, each coding for a particular amino acid, and a gene is a collection of codons lined up in a specific order. That order translates into a chain of amino acids, which forms the basis for a specific protein molecule used by the organism. The four letters of the genetic code can produce a total of 64 possible codons (4^3). However, organisms use only 23 amino acids for protein synthesis, and so most of the amino acids are represented in the code by more than one codon.

However, there are also codons that act as markers to show where one gene starts and ends. ATG is the so-called initiation, or start, codon. Each codon after that is translated into an amino acid (including ATG, which subsequently represents the amino acid methionine). The process continues until the gene presents one of three possible stop codons, which halt the translation process.

Codons…	Code for…	Codons…	Code for…
ATG	START	TTA, TTG, CTT, CTC, CTA, CTG	Leucine
GCT, GCC,GCA,GCG	Alanine	AAA, AAG	Lysine
CGT, CGC, CGA, CGG, AGA, AGG	Arginine	ATG	Methionine
AAT, AAC	Asparagine	TTT, TTC	Phenylalanine
GAT, GAC	Apartic acid	CCT, CCC, CCA, CCG	Proline
TGT, TGC	Cysteine	TCT, TCC, TCA, TCG, AGT, AGC	Serine
CAA, CAG	Glutamine	ACT, ACC, ACA, ACG	Threonine
GAA, GAG	Glutamic acid	TGG	Tryptophan
GGT, GGC, GGA, GGG	Glycine	TAT, TAC	Tyrosine
CAT, CAC	Histidine	GTT, GTC, GTA, GTG	Valine
ATT, ATC, ATA	Isoleucine	TAA, TGA, TAG	STOP

Amino acids

It is frequently said that amino acids are the building blocks of proteins. That is true enough, but in fact only 23 out of the 500-plus known amino acids are used for this purpose (while a handful of others have metabolic roles not linked to proteins). Amino acids are simply a class of organic (carbon-based) chemical compounds. All of them contain a carboxylic acid grouping – the same one that gives vinegar and lemon juice their acidity – plus an amine group, which contains nitrogen.

This nitrogen content is the crucial factor. Nitrogen is very common in the biosphere – nearly 80 per cent of the air is made from it. However, animals are unable to access free nitrogen directly. Instead, they must consume the bodies of other organisms to get the amines they need to build protein. Plants can construct amino acids from nitrate compounds in the soil (hence, why farmers top this up using fertilizers.) However, even they must rely on bacteria to 'fix' nitrogen gas from the atmosphere and make it available to them in nitrate form.

All amino acids have a common structure, with an NH_2 amine group and a COOH carboxylic acid group bonded to a molecule of carbons, R, which can be of any size and shape.

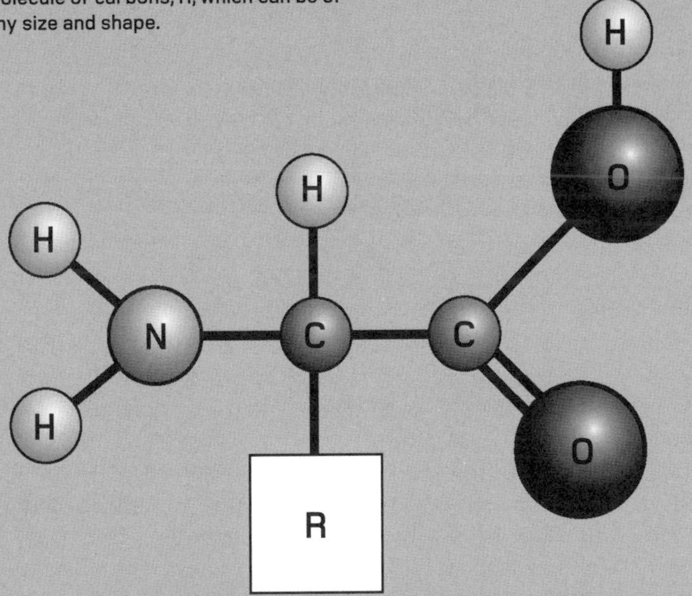

Proteins

More than 100,000 different proteins are used in the human body, all constructed from long chains of amino acids. A single chain of amino acids is called a polypeptide. One cistron (DNA unit or 'gene') carries the code for one polypeptide, and a protein molecule contains at least one of these – many have two or three, each coded for on separate cistrons.

A protein can have anywhere from 400 to 27,000 amino acids inside it. The number of different permutations of acids in molecules of this size is effectively limitless, but the precise ordering encoded in the genes gives every protein a unique shape, which in turn gives it a specific metabolic role. The order of amino acids is a protein's primary structure, while bonds between amino acids at different points on the chain create a twisted secondary structure, which then folds in on itself to make a complex tertiary structure. A single change in a protein's primary structure results in new secondary and tertiary structures, creating a very different molecule.

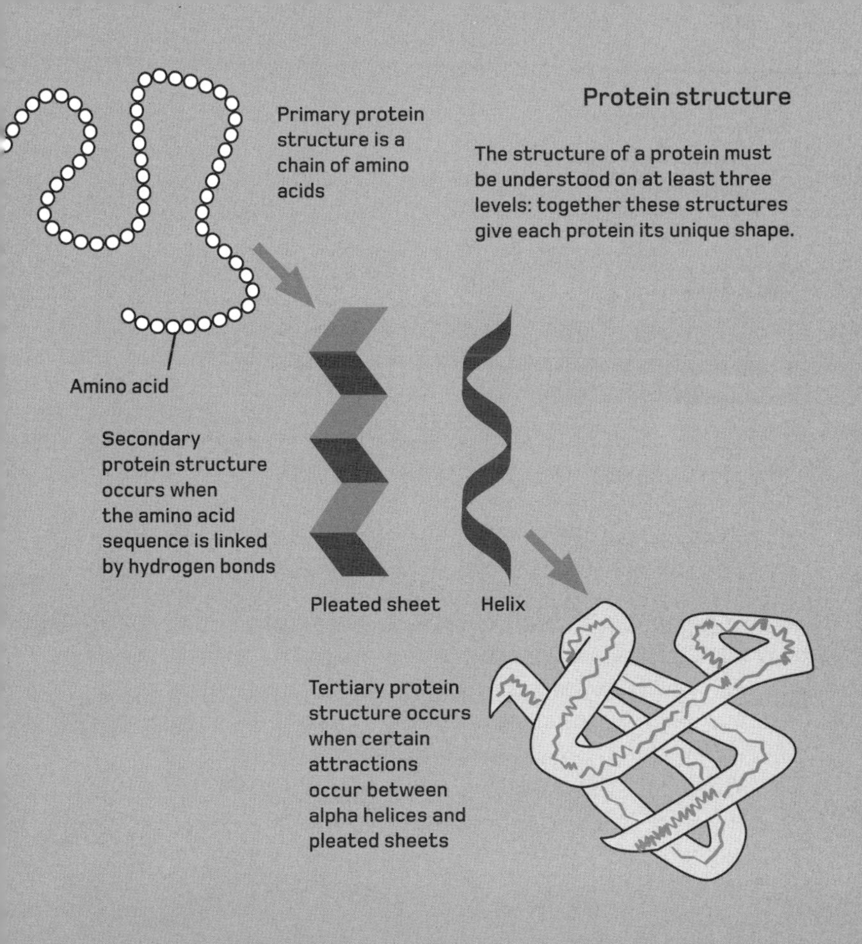

Protein structure
The structure of a protein must be understood on at least three levels: together these structures give each protein its unique shape.
Primary protein structure is a chain of amino acids
Amino acid
Secondary protein structure occurs when the amino acid sequence is linked by hydrogen bonds
Pleated sheet
Helix
Tertiary protein structure occurs when certain attractions occur between alpha helices and pleated sheets

Enzymes

Proteins are commonly seen as synonymous with meat, and particularly muscle. But while it's true that the contraction of muscles is brought about by paired protein molecules pulling against each other, protein has a structural role throughout the animal body. Collagen is the primary example – it forms the foundation layer of skin and other connective tissues that hold the body together.

However, most proteins are used as enzymes. These are the biological equivalent of chemical catalysts, and facilitate the many chemical reactions that are required to keep an organism alive. Enzymes are intimately involved in these reactions, but are not used up by them. Each enzyme has a specific set of abilities, based on its shape, and they are at work both inside the cell and beyond it. For example, the replication of DNA is managed in the nucleus by an enzyme called DNA polymerase. By contrast, amylase is a digestive enzyme, secreted into the mouth and stomach: its role is to break starchy foods into simpler sugars.

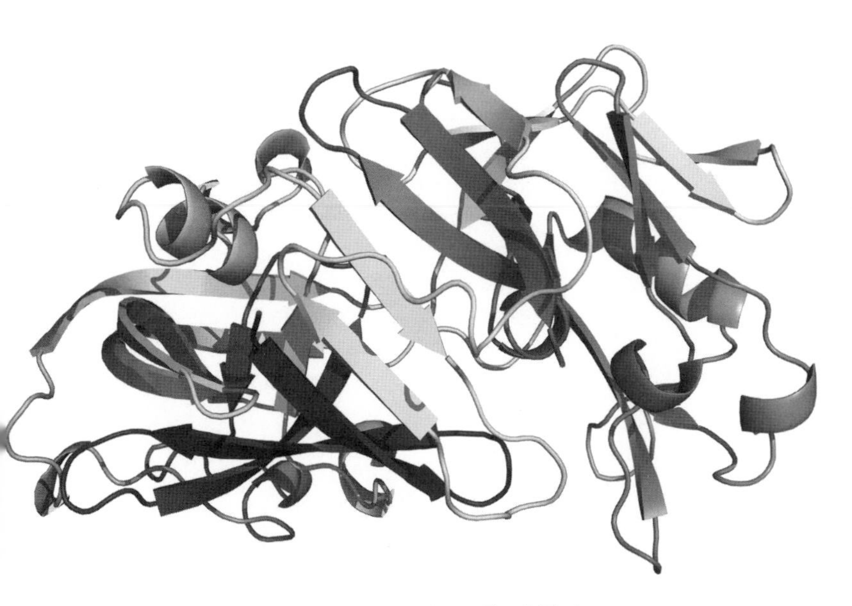

An enzyme's function arises from its shape. The folded molecule is able to bring other substances together to react in a way that would not otherwise happen.

Lock and key theory

An enzyme acts on specific target molecules known as the substrate. This may be one molecule that is split apart by the enzyme's action, or two or more molecules that are joined together. The enzyme is able to make substrates react in ways that would not occur otherwise, because the energy required for the reaction to start spontaneously is too great. So an enzyme is able to manipulate the substrate in such a way that this energy barrier is removed. The best guess at how it does that is the 'lock and key theory'.

In this model, the enzyme is a 'lock', meaning that it has an active site – a location that contacts the substrate – with a specific shape. The substrate is the 'key' that fits into the active site exactly. When fitted together, the enzyme is able to weaken particular bonds in the substrate, pulling some areas apart and bringing others together to bring about the reaction. Often attendant molecules or coenzymes, which are frequently derived from vitamins, are required for a full enzyme function.

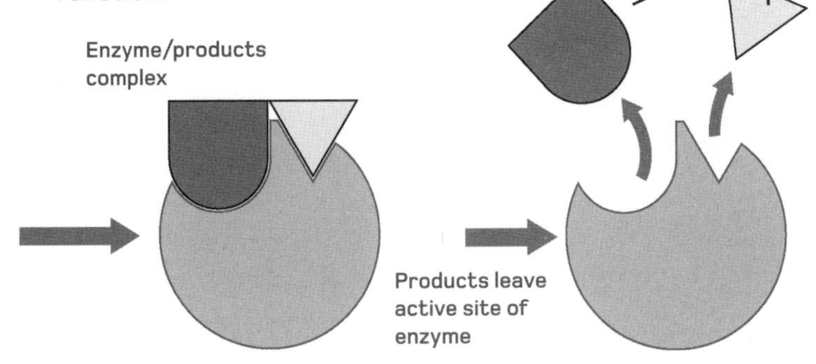

Sex

Considering its significance to life on Earth – it literally made every single human – sexual reproduction remains something of an enigma. The great majority of plants and animals use it to produce young, and on the face of it the advantages are obvious: offspring produced sexually contain a mixture of genes inherited from both parents. If mates are chosen wisely, weaker genes can be paired with better ones, creating a vigorous variety that improves survival chances. But exactly how such a system evolved is a mystery. The first sexual organisms would be easily outcompeted by the extant asexual ones, who passed on all their genes at every generation and were able to reproduce without the need to find a mate.

Nevertheless, sexual reproduction is the dominant strategy among animals and plants. That has meant most life forms are divided into two types, or sexes: male or female. Sexual reproduction requires that the sexes work together to produce young.

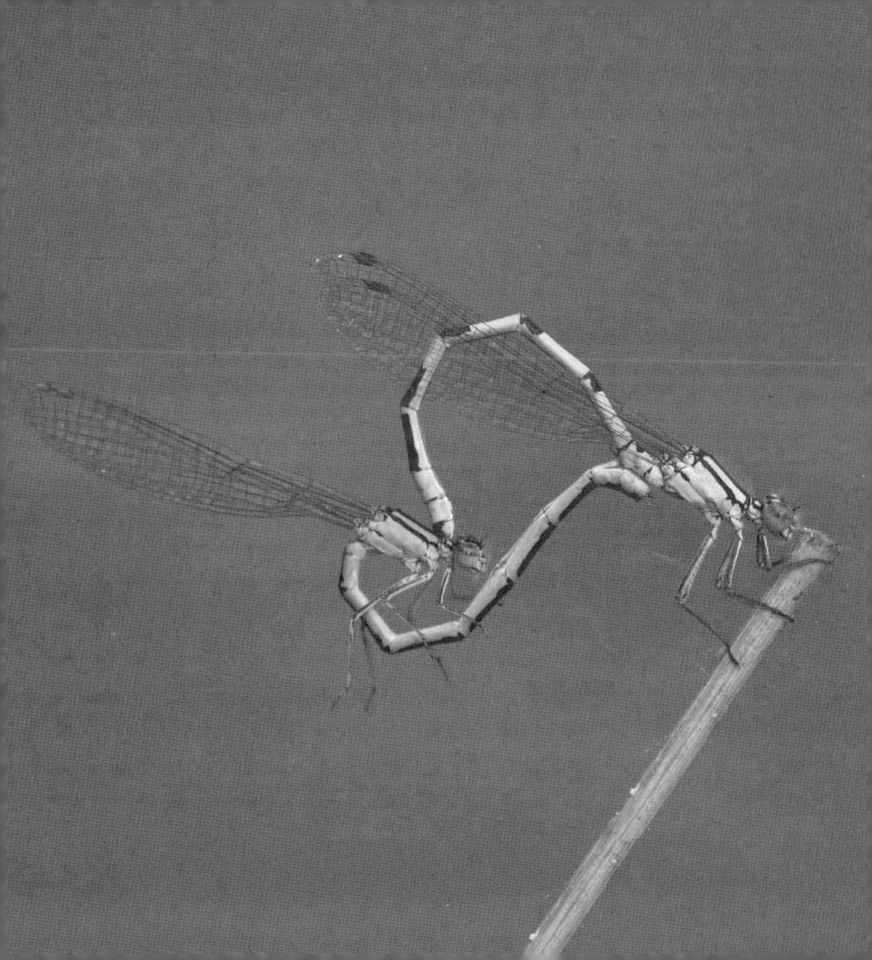

Binary fission

Before the advent of sexual reproduction, all life forms would have reproduced asexually. That means all offspring have just one parent, and all individuals are capable of reproducing by themselves. Asexual reproduction is quick and efficient and allows a single individual alone to populate an empty habitat. The simplest method of asexual reproduction is binary fission – or put more simply, splitting in two.

Only single-celled organisms can breed by binary fission: it is used by all bacteria and other prokaryotes plus many unicellular eukaryotes. The process involves a cell division similar to mitosis (see page 76): The organism's genetic material is duplicated, the cell doubles in volume, and then divides into two new cells. The concept of parent and offspring breaks down at this point. One cell could be regarded as the original, but in general the parent cell is said to produce two daughter cells. In optimal conditions, a bacterium can perform a binary fission every 20 minutes, and so one cell grows into 5 sextillion (5×10^{21}) in 24 hours.

Binary fission is the simplest form of reproduction, used by single-celled organisms that grow in number by dividing in two, to make two identical copies.

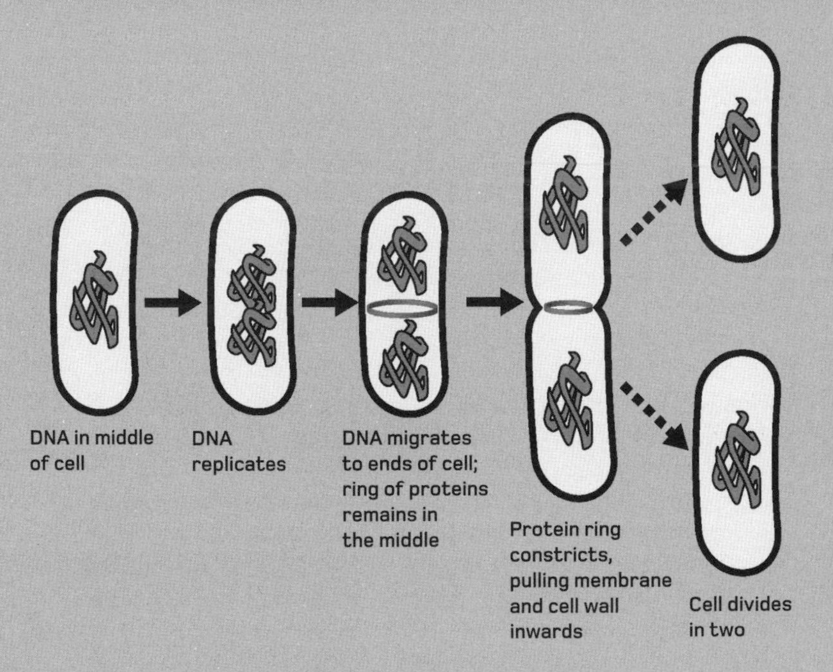

Conjugation

The daughter cells of a binary fission contain 100 per cent of the genes of their parent cell, creating minimal variation between one bacterium and the next. The advantages of rapid reproduction must be weighed against this lack of variety – the population can grow at an exponential rate but any attack may result in a mass die-off of similar proportions: if one bacterium is killed by the threat then so will all the rest. To counter this problem, bacteria have evolved a form of genetic transfer called conjugation.

This involves a donor bacterium transferring DNA to a recipient. Only a small portion of the donor's genome is transferred, in the form of a plasmid, or loop of DNA. The donor connects to a recipient via a pilus, a hairlike extension of the cell membrane that pulls the cells together so they can form a temporary connection. Conjugation only takes place when the recipient does not contain the plasmid already. This ensures that the process always results in the spread of genes.

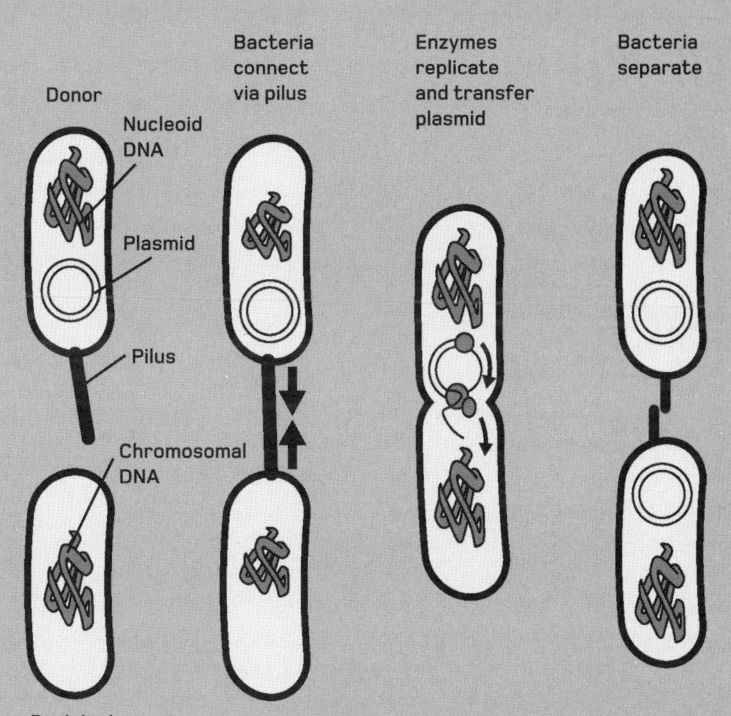

Conjugation is sometimes called bacterial sex. It involves a small ring of DNA being transferred from one cell to an unrelated neighbour.

Budding

While binary fission results in two identical daughters, another form of asexual reproduction produces a clear distinction between parent and daughter. This process, known as budding, is not limited to unicellular organisms: simple multicellular organisms, such as corals, flatworms and sponges do it as well. As the name suggests, budding does not simply involve a parent dividing in two: instead, the offspring develops as an outgrowth or bud on the parent's body.

When it has reached a large enough size to live independently, the bud breaks off. This daughter is smaller than its parent, and will continue to grow to reach a mature size before it starts producing buds of its own. Understanding how animals can grow an entirely new body in this fashion may have implications for using human stem cells (see page 400) to heal injuries. Fragmentation is another alternative form of asexual reproduction, used by worms and some starfish – the parent breaks into several pieces, each of which grows to a full size.

A hydroid, a relative of jellyfish, reproduces by simply growing and releasing a section, or bud, of its body into the water. This then becomes a new individual.

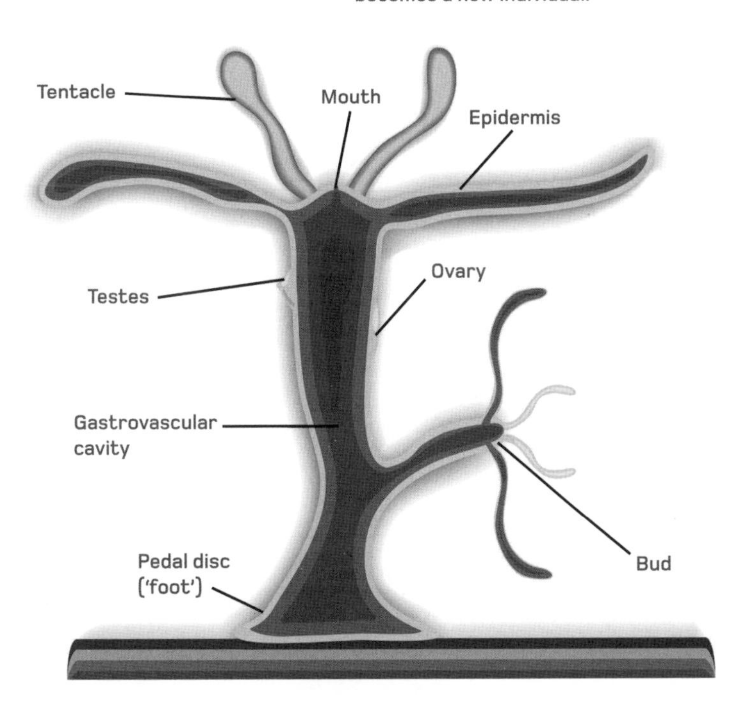

Ploidy

Asexual reproduction involves a simple replication of genetic material that is passed to the next generation in its entirety. Sexual reproduction requires an offspring to receive genetic material from both parents. Mendel's Law of Independent Assortment tells us that all alleles are passed on independently of each other, so this precludes the idea that a male parent provides one half of the alleles while the female provides the other half. Instead, both sexes provide a full set of alleles, and the offspring's cells therefore contain a double set.

This concept is summed up by the principle of 'ploidy'. An asexual organism is monoploid, which means it has one set of alleles in its cells. Sexual organisms are diploid; they have two sets of alleles, but for the purpose of sexual reproduction, the double sets are segregated again into single sets. The result of this segregation is a sex cell, or gamete, containing just one set of alleles. Perhaps confusingly, gametes are sometimes described as haploid, implying half the usual (diploid) number of alleles.

The haploid number is the number of chromosomes needed to carry a full set of genes. Most body cells are diploid, meaning they have two full sets of chromosomes.

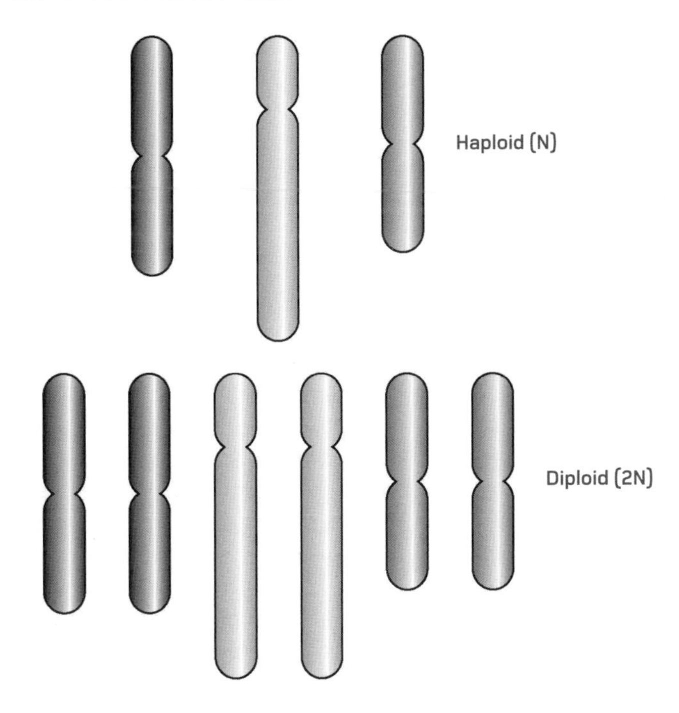

Sperm

The male gamete, or sex cell, is the sperm. Typically, it is a highly mobile cell, propelled by a single flagellum. A surprising number of organisms use this kind of actively swimming 'motile' sex cell – not only animals, but also mosses, ferns and some coniferous plants. Flowering plants and fungi produce non-motile sperm – in the case of plants they are encased in structures like pollen grains. These still move, but must rely on alternative means of transport (see page 164).

The difference between the sexes is encapsulated in a comparison of a sperm with its opposite number, the ovum or egg. A sperm can travel Herculean distances if required. However, it carries only a tiny cargo – a haploid set of genetic material. When it meets the egg, that load is transferred inside, and the sperm's job is complete. This is the defining contribution of the male sex to reproduction, and it means males can produce vast quantities of sex cells at minimal cost in terms of biological resources and energy.

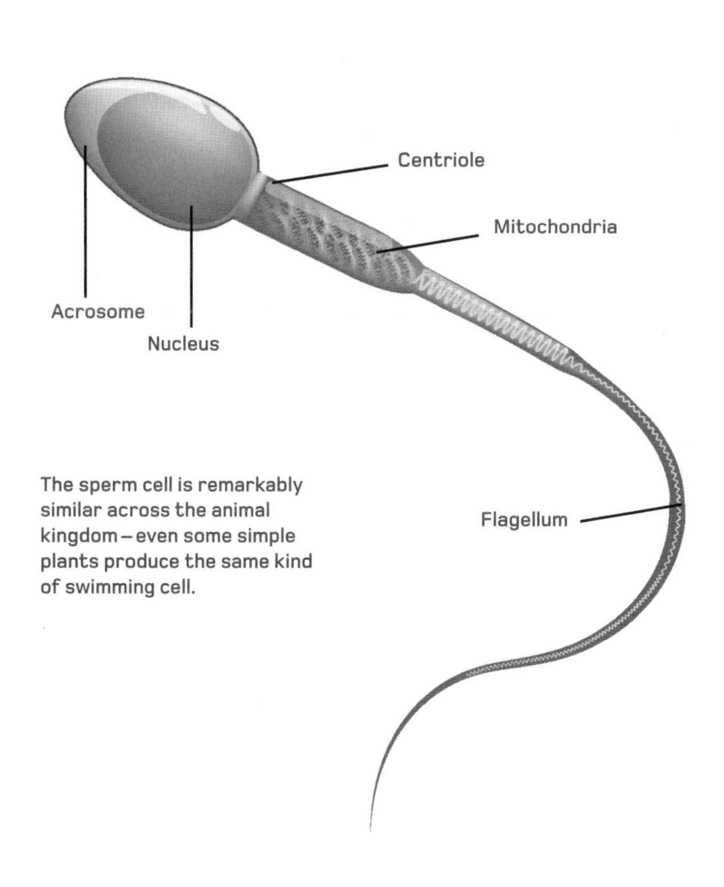

The sperm cell is remarkably similar across the animal kingdom—even some simple plants produce the same kind of swimming cell.

Ovum

Also known as an egg cell, the ovum is the female gamete. It could hardly be more different from the male sperm – a sperm is around 0.05 mm long, including its long tail, while a human ovum is pretty typical at around 0.1 mm wide and just about visible to the naked eye. The cells found in bird and reptile eggs are enormously larger than this.

The greater size indicates the purpose of the ovum. Like the sperm, it is haploid and contains half a full set of genes in its nucleus. The sperm's genetic load passes to the ovum, and, the ovum also carries all the nutrients and cellular equipment required to power the growth of a new individual. This material is stored in the cell's voluminous cytoplasm, termed specifically the ooplasm, but often better known as the egg's yolk. As in any cell, the ooplasm is surrounded by a membrane, but there are further layers around the cell that provide protection and are there to receive the successful sperm – and ensure that no extra rivals get in.

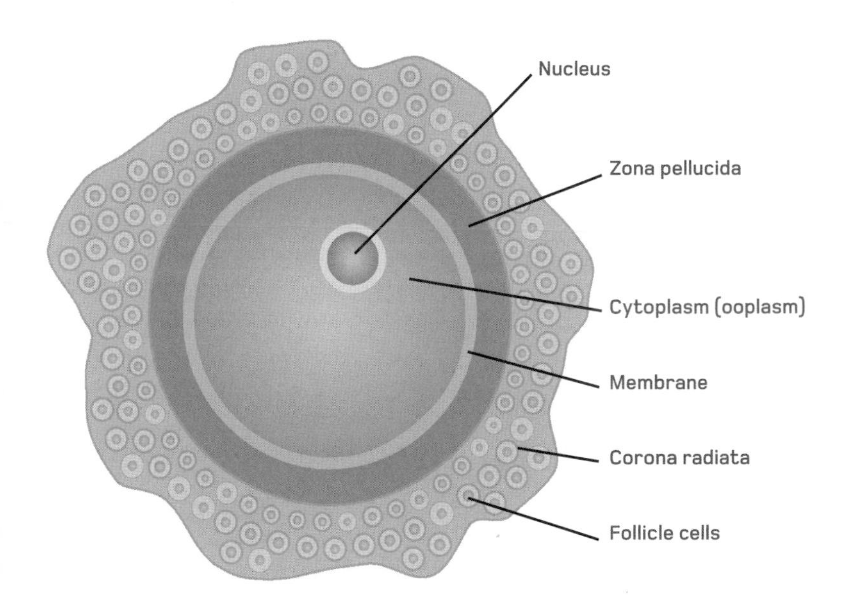

The egg cell, or ovum, contains the material needed to become the first cell of a new organism once it receives chromosomes from a sperm.

Meiosis

Gametes are the only haploid cells in an animal body, and therefore can only be produced by a special kind of cell division called meiosis. This converts one diploid cell into not two, but *four* haploid cells. Meiosis occurs in the gonads, or sex organs. The female gonad is almost universally termed the ovary, while the male one goes by various names (testis, in the case of humans).

The same spindle machinery used in mitosis to separate genetic material (see page 76) is at work in meiosis, but with one crucial difference: meiosis is really *two* divisions. The first division makes two haploid cells. It does this by grouping the chromosomes into homologues – pairs of chromosomes that carry the same alleles, each one originally inherited from either parent. The first division separates the homologous pairs, while the second division pulls the chromosomes (doubled into chromatids) apart as in mitosis. The result is four daughter cells with half the number of chromosomes of the parent cell.

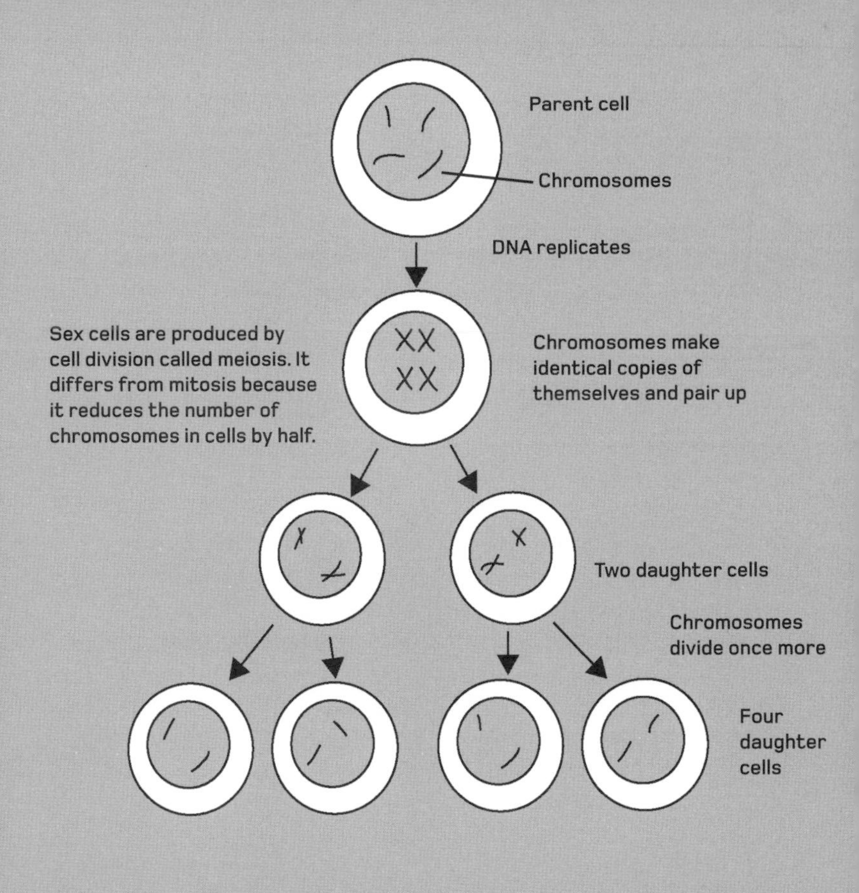
Parent cell
Chromosomes
DNA replicates
Sex cells are produced by cell division called meiosis. It differs from mitosis because it reduces the number of chromosomes in cells by half.
Chromosomes make identical copies of themselves and pair up
Two daughter cells
Chromosomes divide once more
Four daughter cells

Crossing over

Meiosis results in the chromosomes inherited from one parent being thoroughly shuffled with the set inherited from the other. This is done at the level of the chromosome during the first meiotic division. The homologous pairs are split randomly, so paternal and maternal chromosomes can end up together in the cells that result. However, there is a further shuffling of genes that takes place between homologous chromosomes in a process called chromosomal crossover.

Crossover occurs when homologous pairs are lined up ready for the first division of meiosis. At this stage the chromosomes are made up of two identical chromatids, and chromatids from adjacent chromosomes become entwined. Where they cross over, chunks of chromosome are cut and swapped with the neighbour. This results in the chromatids on each chromosome – once identical – now carrying different genes. In the end, the four haploid cells produced by meiosis will contain a unique version of each chromosome.

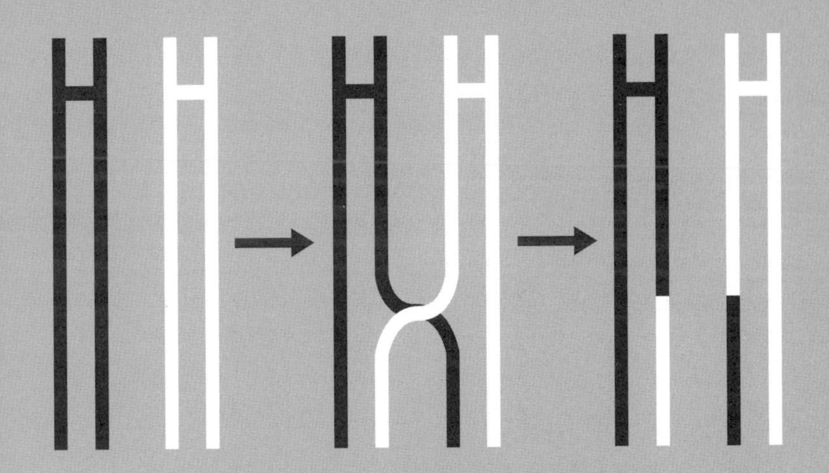

Crossing over during meiosis results in chromosomes from parents mixing their genes to make unique combinations that are then passed on to offspring.

Conception

The formation of gametes is only the first stage in the creation of offspring by sexual reproduction. For a new individual to be formed, a male and female gamete must fuse in a process called either conception or fertilization.

For conception to take place, the gametes need to come together at the same place at the same time. Humans make use of internal fertilization via the tried and tested method of copulation (many other animals do this as well with a great variety of techniques). Fish, frogs and many invertebrates rely on external fertilization, where sperm and eggs are mixed together outside the body. Higher plants have a passive transfer of gametes in the process of pollination (see page 164)

At the cellular level, one sperm fertilizes one egg. The sperm burrows through the ovum's outer layers and gives up its chromosomes, making the cell diploid. It is now a 'zygote', the first cell of a new, unique individual.

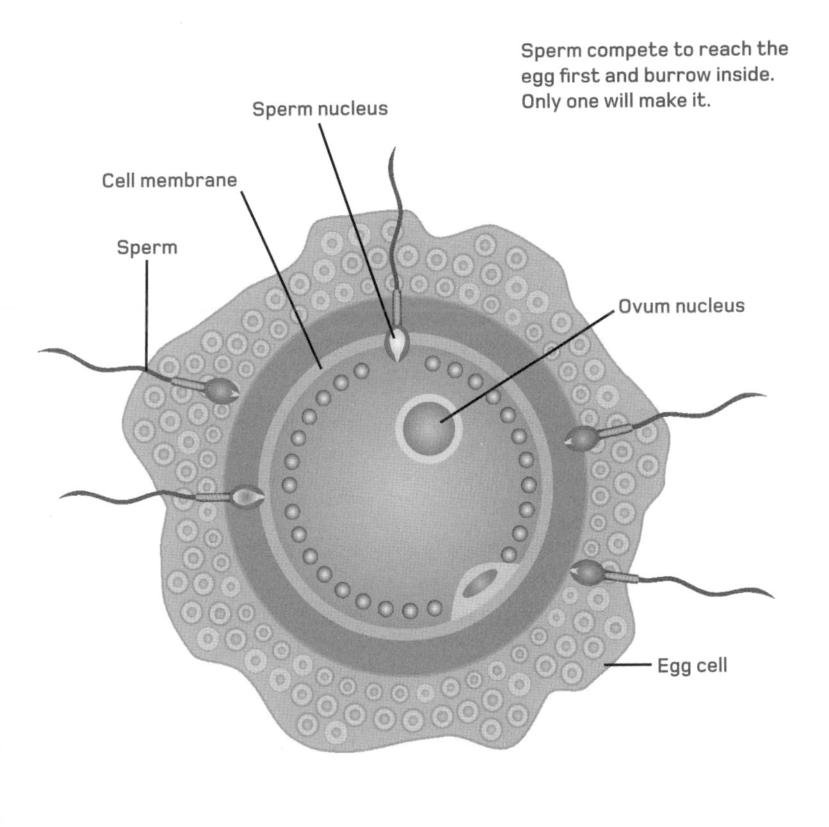
Sperm compete to reach the egg first and burrow inside. Only one will make it.
Sperm nucleus
Cell membrane
Sperm
Ovum nucleus
Egg cell

Embryo

The embryo is the earliest stage of an organism's development. It typically involves a period of rapid growth leading to the organism becoming able to live independently – at which point it hatches out or is born. A different approach is used by plants, where the embryo is contained in the seed. The main growth and development of the plant does not begin until after the seed germinates, sprouting into an independently living individual.

All embryos begin with a single cell, known as the zygote. This is the diploid product of the fusion of two haploid sex cells. Using energy stored in the yolk or ooplasm, the zygote divides by mitosis, rapidly forming a ball of cells. In animals this ball is called the blastula, and from here the cells begin to differentiate into the different layers and tissue types that will make up the eventual animal body. Plant embryos contain an embryonic stem called the hypocotyl, a root or radical, and one or two nutrient-packed embryonic leaves, called cotyledons.

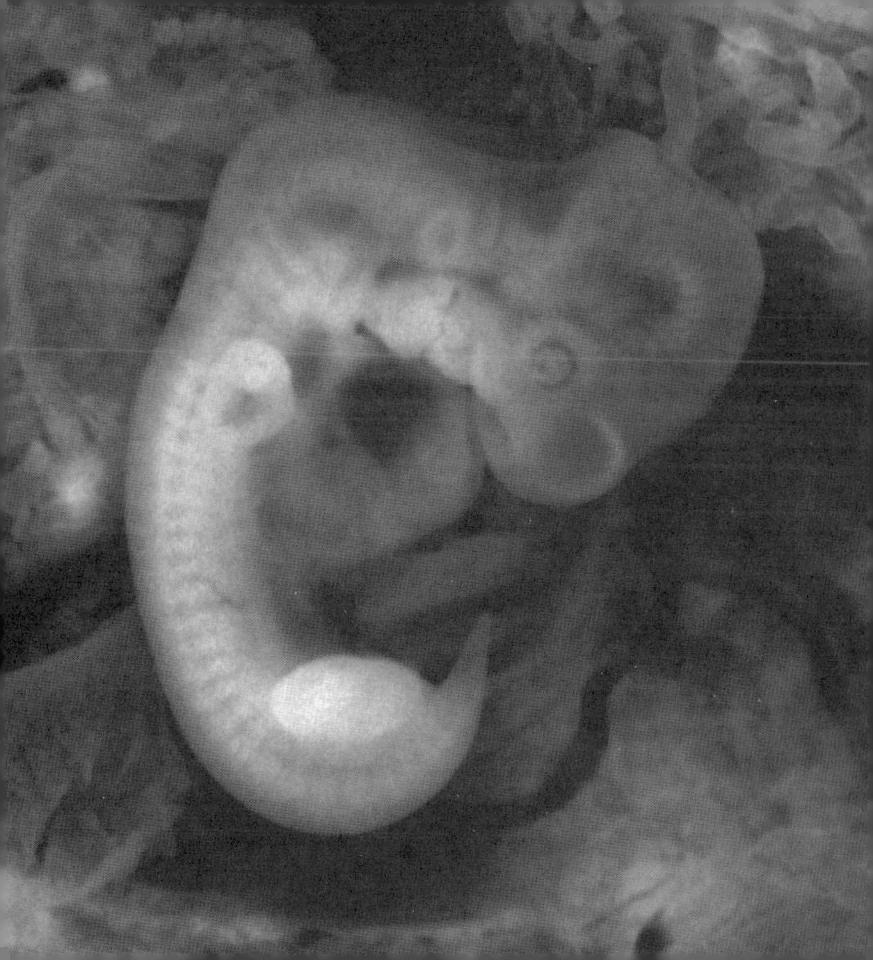

Stem cells

The concept of 'stem cells' is becoming a familiar one, promoted as an exciting new medical tool with the potential to rebuild damaged and diseased body parts. This is possible because all bodies are constructed from stem cells in the first place. Any complex organism is composed of many different cell types that are specialized to perform particular jobs. Once specialized, a cell and its descendants cannot be deployed to another role. Only a stem cell is able to change its function.

The zygote is said to be a 'totipotent' stem cell. That means it is able to specialize into any cell type – and to produce more totipotent stem cells. The embryo grows from these totipotent cells, which specialize through successive levels to produce the many cell types in the body. A fully grown adult body also contains stem cells. These are said to be 'pluripotent', meaning they cannot be used to build a new embryo, but they can develop into almost any cell type already in the body.

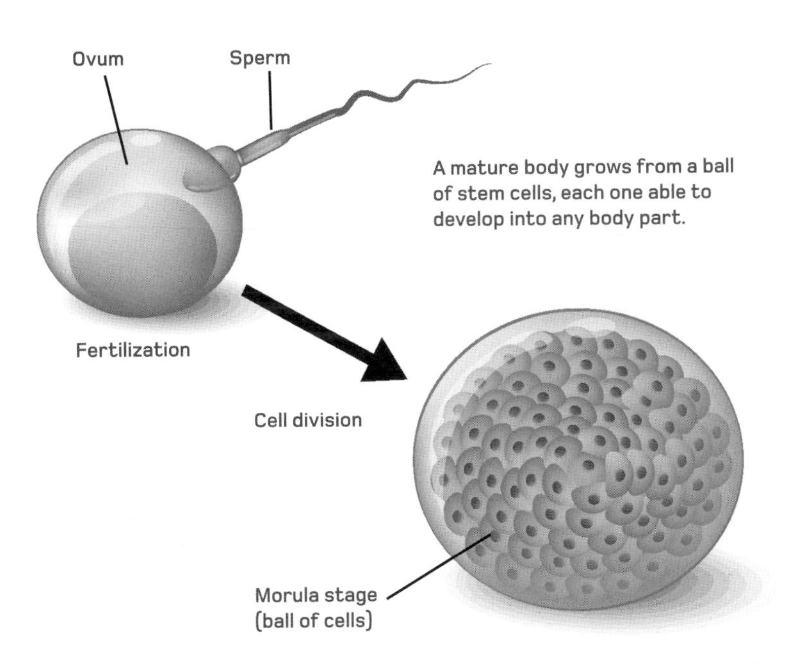
Ovum
Sperm
A mature body grows from a ball of stem cells, each one able to develop into any body part.
Fertilization
Cell division
Morula stage (ball of cells)

Cell differentiation

A multicellular body is a mass of genetically identical cells that are cooperating with each other. Each cell is differentiated in some way, meaning it performs a specific role in the body by deploying a particular set of its genes. There are three main cell types in a body: germ cells, somatic cells and stem cells. Stem cells create the other two types, germ cells develop into gametes, and somatic cells make up everything else.

Somatic cells arise from a cascade of stem cell divisions. A totipotent cell divides into pluripotent cells, which in turn produce 'multipotent' cells. These have the potential to become one of an entire class of related somatic cells – different kinds of blood cell, for example. There may be more stages where the potency of a cell diminishes further, until it arrives at a specific type of somatic cell (a red blood cell, for example). Somatic cells are generally incapable of dividing themselves (liver cells are an exception), and so new ones can only be produced by the action of stem cells.

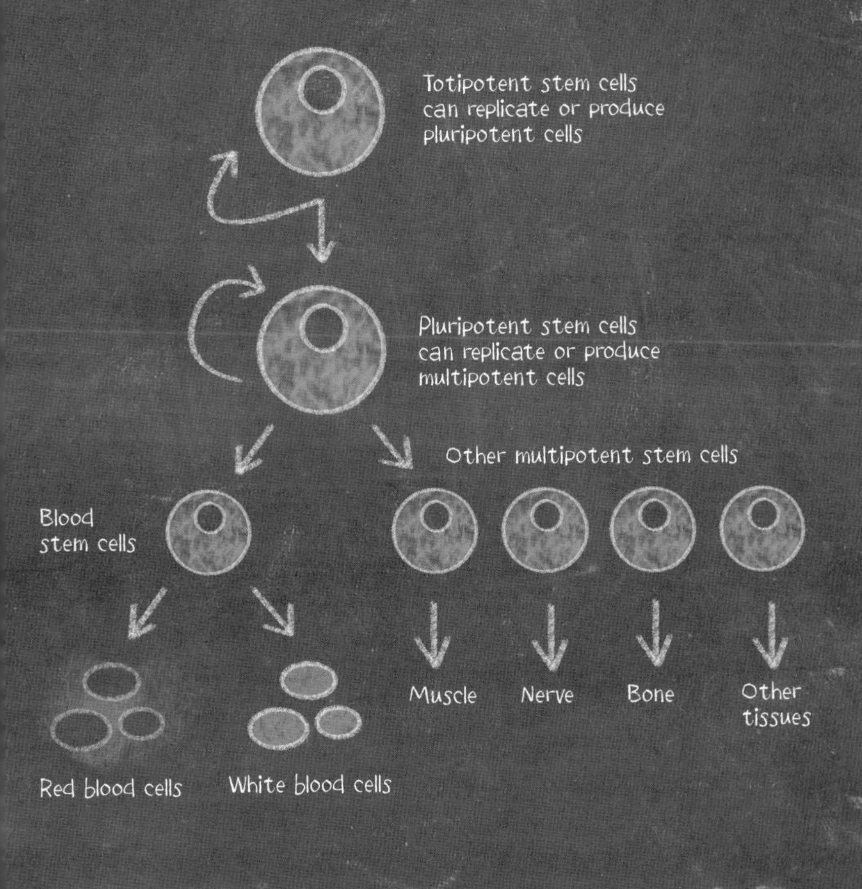
Totipotent stem cells can replicate or produce pluripotent cells
Pluripotent stem cells can replicate or produce multipotent cells
Other multipotent stem cells
Blood stem cells
Muscle
Nerve
Bone
Other tissues
Red blood cells
White blood cells

Tissue

In biological terms, tissues are one way of understanding the different systems at work in a body. A tissue is a group of cells that all originate from the same source – a particular kind of stem cell – and which are all using the same genetic instructions to carry out a particular job in the body. Examples would include the muscles, the lining of the gut and the vessels that run through a plant's stem and leaves.

Ignoring simple organisms like sponges, nearly all animal tissues are derived from three layers of cells that form right at the beginning of an embryo's development. (Plants also have a three-layer development, although it is unrelated.) The ectoderm, or outer layer of cells, develops into nervous tissue, including the brain, skin, teeth, hairs and sweat glands, etc. The mesoderm, or middle layer, becomes the connective tissues, such as bone, blood vessels, cartilage and muscle. Finally, the endoderm, the inner layer, forms the internal organs, such as the lungs, digestive tract and liver.

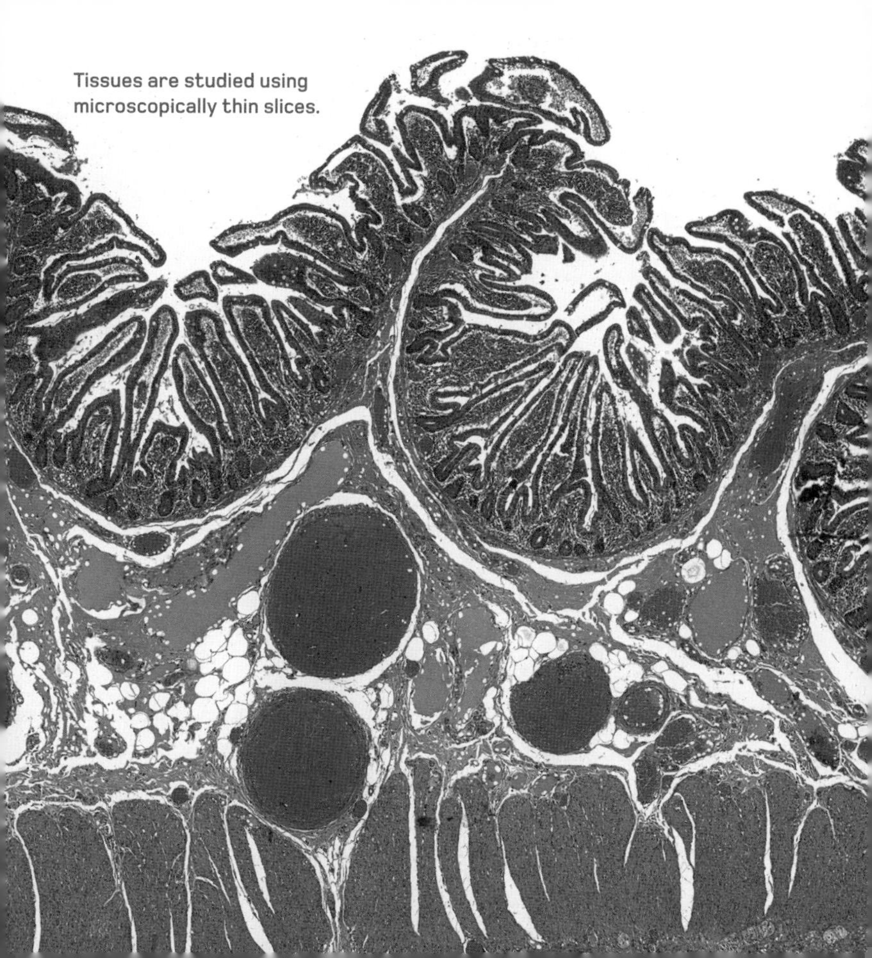

Tissues are studied using microscopically thin slices.

Organs

An organ is the next stage up in complexity from a tissue. Put simply it is a collection of distinct tissues that are massed together to carry out a particular core function in the body. By this definition, a plant's organs would be its roots, stem, leaves and flowers, each comprising a collection of different tissues. In humans, we often refer to the vital organs – the brain, heart, lungs, liver and kidneys – without which sustained life becomes impossible. All animals have some kind of analogous organ performing the same role as each of our vital ones. (For example, fish have gills instead of lungs, while insects excrete not with kidneys but via organs called malpighian tubules.)

As well as the so-called vital organs, a body has many others – the nose, eyes, various glands and, of course, gonads. It is sometimes more helpful to understand individual organs as core components of wider body systems, such as the nervous system, digestive system, circulatory (blood) system and so on.

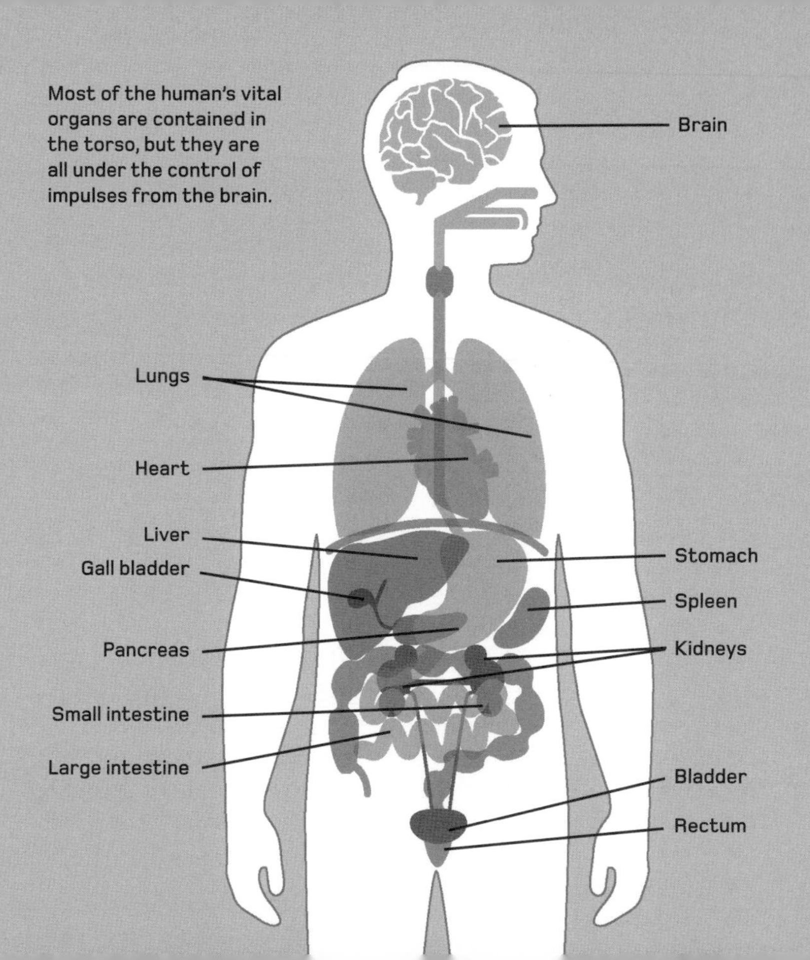
Most of the human's vital organs are contained in the torso, but they are all under the control of impulses from the brain.
Brain
Lungs
Heart
Liver
Gall bladder
Pancreas
Small intestine
Large intestine
Stomach
Spleen
Kidneys
Bladder
Rectum

Oviparity

The life cycles of the majority of animal life involve 'oviparity'. Oviparous organisms develop entirely outside of the body of the mother: the most obvious examples are birds or reptiles that lay their eggs after they have been fertilized internally. The embryo does not really start to grow until it leaves the mother.

A simpler, more primitive version of oviparity is used by fish, frogs and aquatic invertebrates. It involves external fertilization where the female releases her eggs and the male then times the release of his sperm to mix with them as effectively as possible. The fertilized eggs may be simply allowed to float away, or be left stuck in a safe place. However, one or both parents often offer some kind of protection. The external fertilization system makes the male the last parent on the scene – the female can slip away and leave the male 'holding the babies', so to speak. As a result, male fish and frogs are frequently the primary carers of young, the opposite scenario to oviparous organisms that use internal fertilization.

Eggs

The development of an oviparous embryo takes place entirely within a self-contained vessel called an egg. The terminology gets confused here: The egg in question is composed mainly of the constituents of the ovum, but it now contains the zygote. Most reptiles and their evolutionary descendents (including the monotreme mammals, such as the duck-billed platypus) lay what we might easily recognize as an egg – a tough shell containing a yolk. The shell has formed around the egg cell, or ovum, and in effect the whole thing is still one giant cell, vast in comparison to regular body cells. The shell makes the egg waterproof, so it can retain its yolky fuel supply in arid land habitats. However, the shell is permeable to air – the embryo inside needs oxygen to get in – and so shelled eggs cannot survive underwater. The opposite is generally true of unshelled eggs laid by the great majority of oviparous animals. Insects coat their eggs in a waxy sheath to stop them drying out, but in general these eggs have to be underwater, or at least kept moist, for successful development.

Viviparity

The alternative to laying eggs is 'viviparity', where the embryo develops inside the mother and is born without the protection of an egg, at a comparatively advanced stage of development. Mammals are the masters of viviparity, although scorpions, sharks and a few lizards do it as well, though not in exactly the same way.

In fact, there is an important distinction between viviparity and ovoviviparity, a halfway house between egg laying and live birth in which eggs are retained inside the mother for safekeeping but receive no direct nourishment from her. Sometimes the eggs hatch inside the mother, but the young remain inside, and may eat their brothers and sisters. This kind of cannibalistic viviparity is seen in large sharks. Another source of nutrition for the young is oophagy, where they are fed on a supply of infertile eggs produced by the mother's ovaries. The final kind of viviparity involves nutrition supplied from the mother's body via a placenta or similar organ, as seen in humans.

Gestation

The period of time when an embyro is developing inside the mother of a viviparous animal is called the gestation. In mammals, we use the word 'pregnancy', but this describes the state of the mother – gestation refers to the activity of the embryo. Non-mammalian creatures that carry young are seldom called pregnant. Instead, they are said to be 'gravid'.

An embryo gestates inside a space called the uterus. This is generally an enlarged section of the oviduct, the tube connecting the ovary to the genital opening. The embryo inside is supplied with nutrients from the mother. Scorpions do this by developing outgrowths from the uterus, called diverticula, that connect to the mother's intestines, collecting nutrients that secrete through the uterus wall. Most mammals connect the embryo to the mother's blood supply by a placenta. This organ develops from the blastula alongside the embryo. The uterus of marsupials is too small for a working placenta, so their young complete their gestation inside an external pouch.

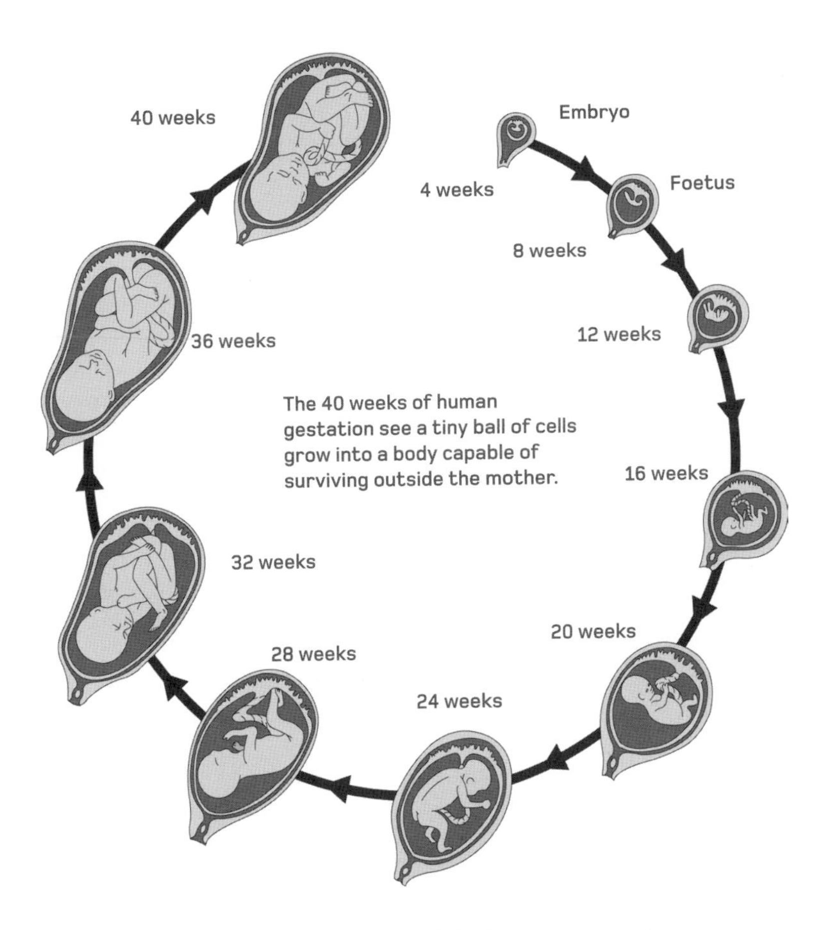
40 weeks
Embryo
4 weeks
Foetus
8 weeks
12 weeks
36 weeks
The 40 weeks of human gestation see a tiny ball of cells grow into a body capable of surviving outside the mother.
16 weeks
32 weeks
20 weeks
28 weeks
24 weeks

Multiple births

A single pregnancy may involve multiple gestations, and result in more than one offspring being born. In smaller mammals this is the norm, with the female releasing several eggs at once from the ovary. Virginia opossums give birth to 50 or more babies at once – although the mother can only suckle a maximum of 13, so most die straight away. The offspring in such a litter are fraternal, meaning they are born at the same time but developed from different zygotes. This means they are no more closely related than brothers and sisters born at a different time. In human terms, two babies born this way are fraternal twins – although they might be brother and sister!

However, it is also possible for multiple births to arise from a single fertilized egg – a single zygote. For example, armadillo mothers habitually produce identical quadruplets from a single zygote that splits, or cleaves, early in its development. When this happens in humans it creates identical twins, which are genetically identical and therefore always have the same sex.

COLD DRINKS
SHAKES

Oestrus cycle

Seen in all female placental mammals, the oestrus cycle's purpose is to prepare the uterus for receiving a fertilized egg and supporting the ensuing embryo. In the first week of development, the zygote and blastula float inside the uterus. After that period, having reached a larger size, the ball of cells implants in the wall of the uterus, where it builds a placental connection to the mother's blood supply.

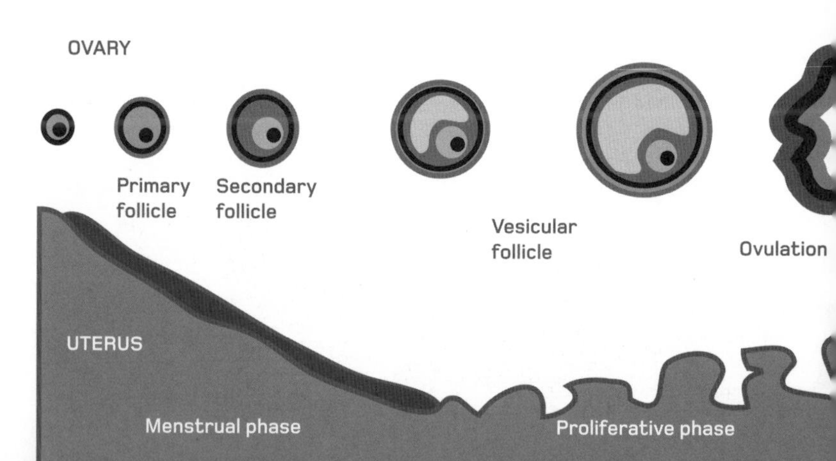

The oestrus cycle ensures that the egg ripens and emerges from the ovary – an event called ovulation – at just the right time so that when it arrives in the uterus, the lining has thickened up ready to receive it. If the egg has met a sperm on its journey to the uterus, the resulting zygote will pause the cycle and keep everything ready for the arrival of the embryo. If fertilization does not take place, the egg decays and the lining of the uterus is shed in a period of menstrual bleeding. Then the cycle begins again with the ovary ripening a new egg and the uterine lining thickening once more.

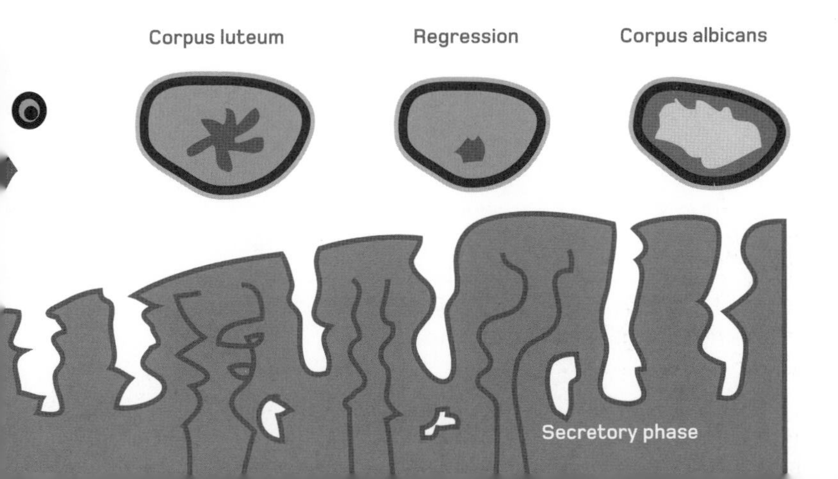

Parthenogenesis

Deriving from the Greek for 'virgin birth', parthenogenesis is a form of asexual reproduction seen in plants, many invertebrates, fish, amphibians and reptiles. A few freak occurrences have been noted in birds but it has never been recorded in mammals. Parthenogenisis uses the mechanism of sexual reproduction, but the young are produced without the need to fertilize the eggs. This is possible when the meiosis process that produces eggs (see page 132) only produces two daughter cells – both diploid. The precise steps vary, but at some point pairs of haploid cells merge back into a single diploid one.

Some species can *only* reproduce by parthenogenesis. All members of the species are necessarily female. However, other species breed parthenogenetically to exploit good conditions, but revert to sexual reproduction at other times. In some systems, a male's sperm (or pollen) is still needed to stimulate the female's egg, but the sperm's genes are not passed on.

An aphid gives birth to a daughter, who already has more identical granddaughters developing inside her body.

Hermaphrodites

There are sometimes misconceptions about the word hermaphrodite, and many people think that hermaphrodites can reproduce without the need for sex. This confusion probably stems from the fact that some hermaphrodites are simply able to have sex with themselves. An hermaphrodite is an organism that has both male and female gonads. This is largely the norm for flowering plants but is also the case for some animals.

Snails, slugs and earthworms are examples of *simultaneous* hermaphrodites – meaning they have both sex organs at the same time, and some (but not all) are able to fertilize their own eggs with their own sperm. Other animals are *sequential* hermaphrodites, starting out life as one sex and changing to another as they get older and bigger. Clownfish (Amphiprioninae family) begin life as males and become females in later life – which may be why *Finding Nemo II* has never been commissioned.

A pair of leopard slugs copulating by twisting their penises together.

Pollination

Pollination is the method of gamete transfer used by higher plants – conifers and flowering plants, which all produce seeds. The male sex cells are housed inside a pollen grain, and it is this body that makes the journey from one plant to the next. Conifers rely on the wind to blow their microscopic pollen grains out of their male cones and into a receptive female cone nearby. Many flowering plants, such as grass and oak trees, rely on wind pollination: their flowers are frequently long, wispy and inconspicuous, built to catch the breeze not the eye. Flowers that rely on insects or other animals to transfer pollen are bright, scented and laden with nectar to attract pollinators.

Once it reaches another flower, the pollen is collected on a tall, sticky receptor called the stigma. To gets its sex cells into the ova, the pollen burrows down to reach the ovary. After fertilization, each ovum grows into an embryo housed within a seed casing, and usually some kind of surrounding fruit that develops from the ovary and remnants of the flower.

Alternation of generations

In animals, the only cells that are haploid (with half the normal compliment of chromosomes) are the reproductive gametes, but the same is not true of plants. Instead, plant bodies alternate between a diploid body and a haploid one in a phenomenon known as the 'alternation of generations'. The diploid form is known as the sporophyte, while the haploid structure is the gametophyte.

In higher plants, gametophytes are small and totally dependent on the sporophyte: the male pollen grain and the female ovule (an egg container deep within the flower). But in 'lower' plants, such as moss and ferns, the two generations form larger structures. In mosses the gametophyte is the dominant form, while in ferns – the ancestors of 'higher' plants – the sporophyte is the main structure. Meiosis within the sporophyte (see page 132) releases haploid spores that grow into a gametophyte, within which sex cells – sperm and eggs – are produced. In heavy rains, the sperm can then swim to neighbouring plants, fertilizing eggs that grow into the next generation of sporophytes.

The familiar shape of ferns is just one of two body forms grown by these plants.

Evolution

The idea that organisms are able to change, or evolve, from one form to another over time is most frequently associated with Charles Darwin. His ideas have indeed won the arguments over evolution, but they were by no means the first. The Greek philosopher Aristotle believed that all natural things strove to fulfil a role in the universe, and this sowed the seed of an idea that living things were able to change their form in pursuit of that goal. As modern scientific methods developed in the 18th century, two opposing views took shape among the naturalists who catalogued living things. Some said that every organism belonged to a certain type of species, and that this was an unchanging aspect of nature. Others pointed to the vast age of Earth (which was becoming better understood by this time) as evidence that today's life forms could have lived in other ways in the past. Erasmus Darwin (grandfather of Charles – see page 176) suggested that large animals had all descended from a microscopic ancestor. The challenge was to find a mechanism by which this could have happened.

Birds, dinosaurs and reptiles all share a common ancestry, but owe their different forms to a variety of evolutionary pressures and processes.

Spontaneous generation

Inherent to early ideas of how organisms evolved was the idea that new life was constantly forming out of non-living things. This concept of 'spontaneous generation' stated that microorganisms (still to be studied in any great detail) arose from the putrid remains of dead organisms and their waste. This was still many decades before cell theory was formulated (see page 72), and so spontaneous generation was still seen as the best explanation for observations that saw moulds and other germs appear as if from nowhere on rotting matter. Even relatively complex life, such as fly maggots and beetle grubs that wriggled out of dung were supposed to have formed directly from inorganic material. Spontaneous generation was thought to provide the source material for evolution, as each primitive life form worked to move up the biological scale. This 'teleological imperative', in which organisms actively sought to improve, came straight out of Aristotelian philosophy and was seen as the driving force behind evolution. It is an idea that is hard to shake even today.

Acquired characteristics

The modern science of biology grew out of the work of naturalists. These were 18th-century nature lovers – frequently men of the cloth – who began to document and above all catalogue the natural world. Each organism fell into a specific type, or species, whose members all shared a particular set of characteristics. Such 'specific' characteristics were deemed to be unchanging, a summation of the unique essence of that organism. This concept was hard to argue against in the years before inheritance was understood.

So when it came to looking for a way that species could differ from each other and effect an evolutionary change, naturalists turned to 'acquired characteristics' rather than specific ones. These are changes the body undergoes in response to its activity and environment – anything from a bodybuilder's muscles or the way skin on the hands becomes thicker with manual labour. Early theories of evolution suggested such acquired traits could be inherited, making it possible for species to evolve.

Lamarckism

The first fully conceived theory of evolution was put forward by French naturalist Jean-Baptiste Lamarck in 1809. He used the term transmutation to describe the way a species could change over many generations. The theory was superseded by Darwinism, but its central idea is now being revived with the discovery of epigenetics (see page 354).

Lamarck believed that evolution had a direction: nature strove to improve and progress towards more complex forms. However, he also introduced the idea that evolution worked to adapt an organism so that it was better suited to the environment than its forebears. The agents of change in 'Larmarckism' were characteristics acquired in life. According to the most famous erroneous example, each generation of giraffe reached up to the freshest leaves, stretching their necks. Their offspring inherited the longer neck, and continued the process, resulting in ever taller generations. Today's giraffes are deemed to be 'tall enough', and so the upward trend has ceased.

Charles Darwin

Charles Darwin (1809–82) is among the most famous scientists of all time, and his theory of evolution by natural selection marks one of the greatest shifts in scientific thinking. Darwin was somewhat reticent about his role as scientific behemoth, and mostly left others to support his theory against those with opposing beliefs – church leaders among them.

Darwin was born in Shropshire – his father was a doctor and his mother an heiress to the Wedgwood porcelain fortune. Charles initially studied medicine at Edinburgh but did not thrive, and so his father moved him to Cambridge, ironically in preparation for taking holy orders. There, Darwin further developed his enthusiasm for natural history, making detailed studies of beetles found in the surrounding countryside. He also studied the work of William Paley, a theologian who had used natural history as evidence of the work of God. Upon graduation, Darwin's eclectic education left him well placed to think the unthinkable.

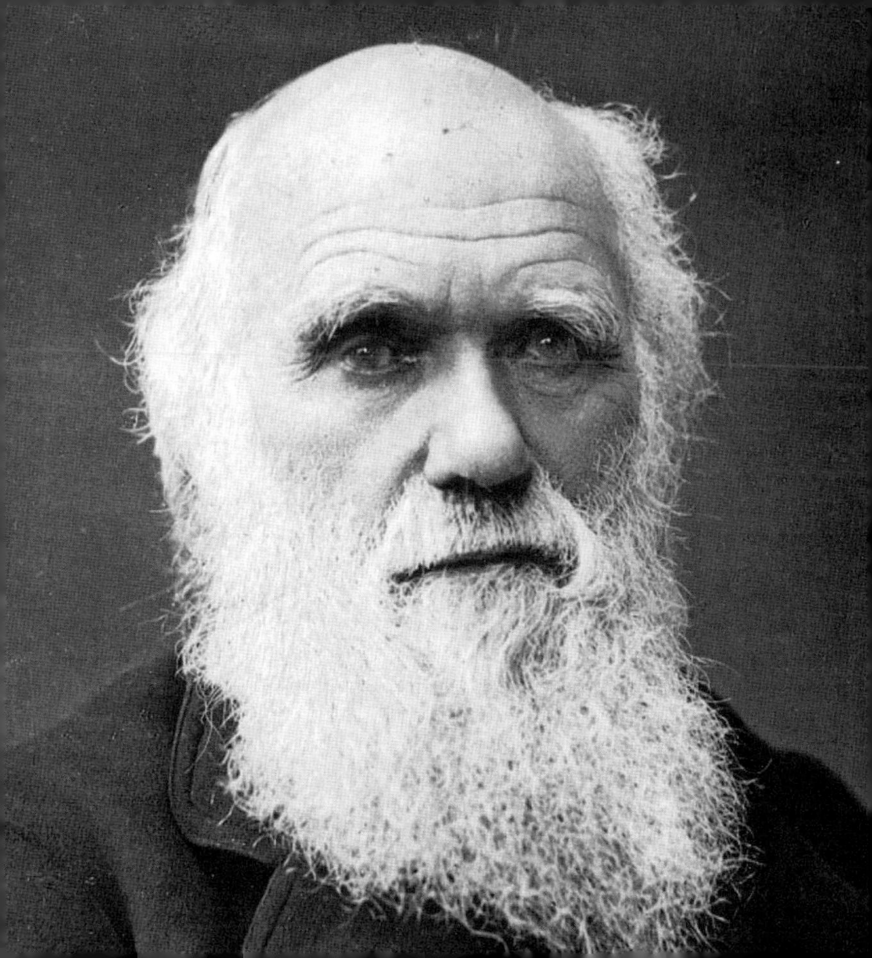

Voyage of the *Beagle*

Many of the most powerful observations that caused Charles Darwin to form his theory of evolution took place during a circumnavigation of the globe aboard HMS *Beagle*. The *Beagle*'s mission was a peaceful one, with orders to survey the coast of South America and Pacific islands. Captain Robert FitzRoy (later founder of the Met Office, Britain's national weather forecaster) called for a scientific-minded civilian to join the crew as his companion. Darwin, who had graduated a few months before, accepted the offer – despite the fact that he would have to pay his own way.

The voyage took Darwin to Africa, South America, New Zealand, Australia and numerous islands, including the Galápagos on the mid-Pacific equator. The journey lasted five years, most of which Darwin spent ashore, collecting and comparing the organisms he found from place to place. The similarities he found between apparently unrelated species on separate continents were the starting point for his theory of evolution.

Alfred Russel Wallace

Darwin was not the only mid-19th century naturalist thinking about how evolution could shape the bodies of animals and plants. While Darwin lived in relative solitude pondering the immensity of his theory in private, Welsh explorer Alfred Russel Wallace (1823–1913) was making his own journey of discovery through the islands of Malaysia and Indonesia. This region of the world marks a boundary between many animals that were ancestral to ancient Australia on one side, and those of Asia on the other. Today, that boundary, running through the Celebes Sea and Lombok Strait, is known as the Wallace Line.

Wallace saw how the most closely related species were found in neighbouring areas, and the differences between them appeared to follow a gradual sequence, as if each species arose from its neighbour. Wallace wrote to Darwin in the 1850s asking for comments on his ideas about how this evolution occurred. As a result, Darwin was spurred into finally going public with his own long-gestating theory.

On the Origin of Species

Upon his return to England from HMS *Beagle*, Charles Darwin married his cousin Emma Wedgwood and settled down to a comfortable life in the country. However, the deaths of three of their ten children in infancy weighed heavily on Darwin, as did the immense implications of the theory of evolution, which he developed over many years. Only a few colleagues got to hear about it, but Darwin planned to eventually present it in a huge opus entitled *Natural Selection*.

In 1858, however, Darwin received a letter from Alfred Russel Wallace outlining a similar theory of evolution. The pair presented joint papers to the Linnaean Society that same year, while Darwin paused *Natural Selection* to dash off a shorter work on his ideas. The result was *On the Origin of Species,* first published in 1859. Amidst copious examples, it explained how all organisms – including humans – are related to a common ancestor in the distant past. Few other books have had such a dramatic effect on the way humanity views itself.

ON

THE ORIGIN OF SPECIES

BY MEANS OF NATURAL SELECTION,

OR THE

PRESERVATION OF FAVOURED RACES IN THE STRUGGLE FOR LIFE.

By CHARLES DARWIN, M.A.,

FELLOW OF THE ROYAL, GEOLOGICAL, LINNÆAN, ETC., SOCIETIES;

AUTHOR OF 'JOURNAL OF RESEARCHES DURING H. M. S. BEAGLE'S VOYAGE ROUND THE WORLD.'

Natural selection

Charles Darwin's theory of evolution invokes the principle of 'natural selection' – an idea with one of its roots in the 1798 *Essay on the Principle of Population*, written by Thomas Malthus (1766–1834) and read by both Darwin and Wallace. Malthus's work warned that growth in the human population was destined to outstrip the ability to grow food, leading to global famine.

Technological advances have prevented this 'Malthusian' catastrophe so far, but to Darwin's naturalist mind it posed the question of how non-human populations were controlled. He reasoned that a wild population had a finite set of resources available to them – food, space, etc – which could only support a finite population. Only some of the population would live, and the rest would die, but the battle for survival would not be random. Nature selected the winners: those that were best able to command the resources they needed would survive, and those unable to do so died. Darwin's masterstroke was to recognize the power of this 'natural selection' to create changes in species.

Competition

To paraphrase the philosopher Thomas Hobbes (1588–1679), life is 'nasty, brutish and short'. This is especially so for populations of wildlife, where a long life followed by a death from old age is a rarity indeed. While earlier theories of evolution supposed that the process was underwritten by some supernatural goal of improvement, Darwin saw that the only thing needed to power evolution was the competition for survival.

All life is competing to survive, battling for a supply of energy, nutrition, oxygen and water – and the space to use it. The strongest competition of all is between members of the same species, which share the same requirements and use the same means to achieve them. In addition, the drive for survival is only a means to an end. The purpose of survival is to reproduce, and individuals compete to maximize their opportunities to do so. The natural selection of competition not only acts to kill weaker individuals, but also prevents them from breeding, blocking them from passing on their genes to the next generation.

Variation

Natural selection needs something to work with: if a population of animals were all identical then none would have an advantage over the others. However, nature is not like that – every population contains a degree of variation, and it is these differences that can make one individual a success and another a failure by comparison.

Lamarck (see page 174) suggested that the long necks of giraffes were due to stretching to reach leaves and became incrementally taller every generation (though how exactly, he could not say). Darwin's explanation chimed better with the known facts. Some giraffes are taller than others; their height gives them an advantage, so they eat more – and breed more – than their smaller neighbours. Short giraffes are more likely to starve and not have young. Darwin understood that tall giraffes have tall offspring, so natural selection results in more tall giraffes being born – and giraffes as a species evolve to be taller. But they are never all *identically* tall; there is always some variation.

Mutation

Darwin understood that for his theory to work, offspring would have to inherit some kind of 'genetic' material from their parents. This material was the means by which the advantages of the parents – those traits selected by nature – would be passed to their young. Thanks to the discovery of DNA and the central dogma (see pages 78 and 98), we now understand a lot more about what that genetic material is and how it works. This also shows us where the variation that feeds evolution ultimately comes from: a population's variation is due to the different alleles in its gene pool. These alternative versions of genes arise randomly due to mutations – errors made during DNA replication. Without such mistakes life would not have evolved at all. If a mutation occurs in an intron, it has no impact. If it appears in an exon, it will alter the structure of the protein coded by the gene. The chances are that this will create a disadvantage, and natural selection will soon wipe it from the gene pool. But occasionally a mutation creates a new kind of advantage – and evolution occurs.

Jacob sheep have four horns, not the usual two, thanks to a genetic mutation.

Survival of the fittest

Although he did not coin it, Charles Darwin readily adopted the term 'survival of the fittest' as a description for his theory of evolution by natural selection. In this context, the term 'fitness' sums up the balance of advantages and disadvantages inherited by an individual. If advantageous traits outweigh the deleterious ones, an individual is 'fit', and would succeed in the battle for survival against less fit competitors. Natural selection ensures that the fittest survive and have the most offspring. Those offspring are also likely to be fit, having inherited advantageous traits – or genes – from their parents.

If a mutant allele arises that provides an advantage over earlier forms, its carrier will be fitter than his or her neighbours. Over the generations, natural selection will result in this mutant gene spreading through the population, while less fit alleles become rare or disappear. The change is small, perhaps imperceptible, but given a great expanse of time and many generations, tiny accrued changes like this can alter species entirely.

Adaptation

Evolution is in some ways a refinement process. Natural selection filters out the unfit genes and ensures that the population as a whole is better suited for survival. However, there is another side to the equation. An individual's fitness can only be measured by the environment it finds itself in. A trout is well suited to life in a river, but it cannot compete among a herd of camels crossing a desert (or vice versa).

The environment in which a population of organisms finds itself is not constant. It can change its character, sometimes very quickly, and this throws up new challenges for survival. Any change will alter an individual's fitness – the genes that once brought success are no longer enough. Natural selection simply carries on, promoting different alleles that provide an advantage in the new conditions. The result is that the organisms can adapt to their new habitat. It is evolution's ability to create adaptations for different environments that has driven life divergence into a multitude of species.

The dark form of peppered moths has become more common than the pale variety as the species has adapted to hiding on trees blackened by industrial pollution.

Speciation

Natural selection moulds organisms to their environments. Over millions of years and many small changes, a group of animals recognized as belonging to one species can change so much that they form an entirely new group. This process of change is called speciation, and there are two main ways it can occur.

The simplest mechanism involves a single-species population becoming divided by a physical barrier. Perhaps an exceptional summer has cleared ice from an alpine pass allowing a herd of goats to pass into a neighbouring valley – but the ice returns to block the route. The two groups of goats now live in different environments, with different foods and predators. As a result they evolve in different ways and become separate species. The second form of speciation occurs within a population: a mutant goat is able to stomach foods that are toxic to the rest of the herd, and so a subpopulation of mutants develops to exploit a different food source to the others. They stop breeding with the non-mutant herd, splitting the population into two species.

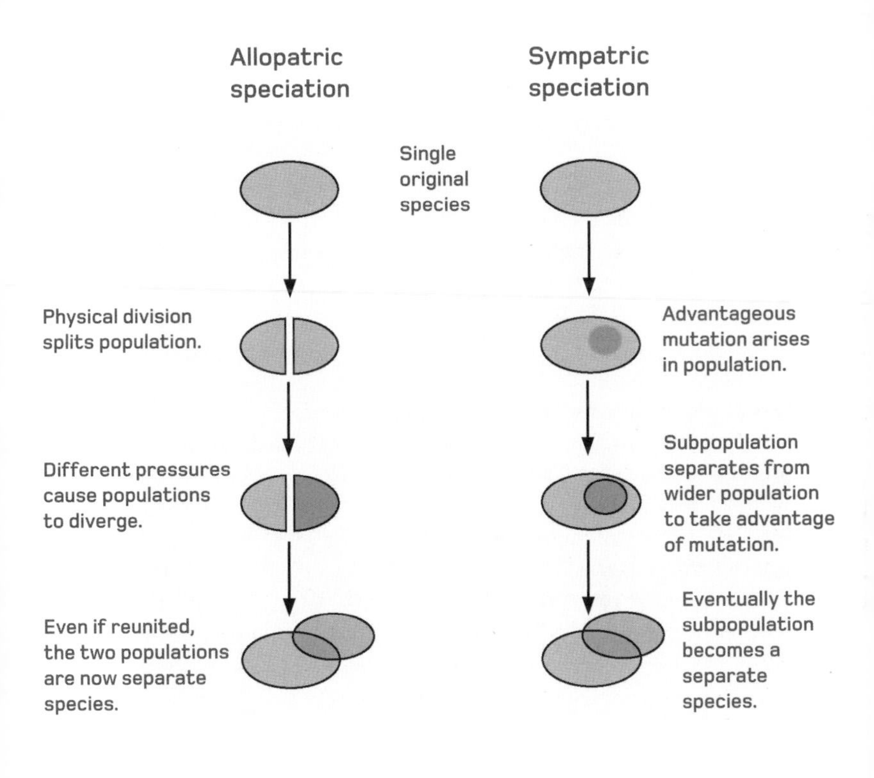
Allopatric speciation
Sympatric speciation
Single original species
Physical division splits population.
Advantageous mutation arises in population.
Different pressures cause populations to diverge.
Subpopulation separates from wider population to take advantage of mutation.
Even if reunited, the two populations are now separate species.
Eventually the subpopulation becomes a separate species.

Extinction

Extinction is perhaps well understood because of the dinosaurs and other fascinating species that are known to have lived in the past. Until the dodo of Mauritius was hunted to extinction around 1688, the idea that a species could die out was barely considered. In 1813, French anatomist Georges Cuvier (1769–1832) revealed that extinction was not just an unnatural act committed by humans. He showed that the fossil remains of mastodons were not the same species as living elephants: as time passes, species become extinct and are replaced by new ones.

According to an oft-quoted statistic, some 99.9 per cent of species that have ever evolved are now extinct, but this requires a little clarification. The dodo and the mastodon are truly extinct, meaning none of their species survive, and neither do any 'daughter' species that evolved from them. However, on a larger scale we can say that dinosaurs are only *pseudoextinct*: today's extant bird species evolved from dinosaurs, and so still carry at least some dinosaur genes.

Convergent evolution

One of the strongest indicators of natural selection at work is convergent evolution. This is the observation that animals with very different ancestries tend to evolve in similar ways when they adapt to the same environment. A good example is the convergent evolution of large pelagic predators – animals that hunt in the open oceans. Sharks are the most ancient: they typically have a streamlined body with fins for stabilizing the body and a wide tail for propulsion. The icthyosaurs, marine reptiles that hunted in the oceans until about 90 million years ago, had the same body shape with fins and tail matching a shark's. Today, dolphins occupy a similar place in the environment – known as an ecological niche – and they too have a similar body plan. These fish, reptile and mammal species all evolved independently but ended up looking very much the same, due to a phenomenon called selection pressure. The blind hand of natural selection tends to push evolution in the same direction, so unrelated organisms develop the same adaptations to survive in a particular niche.

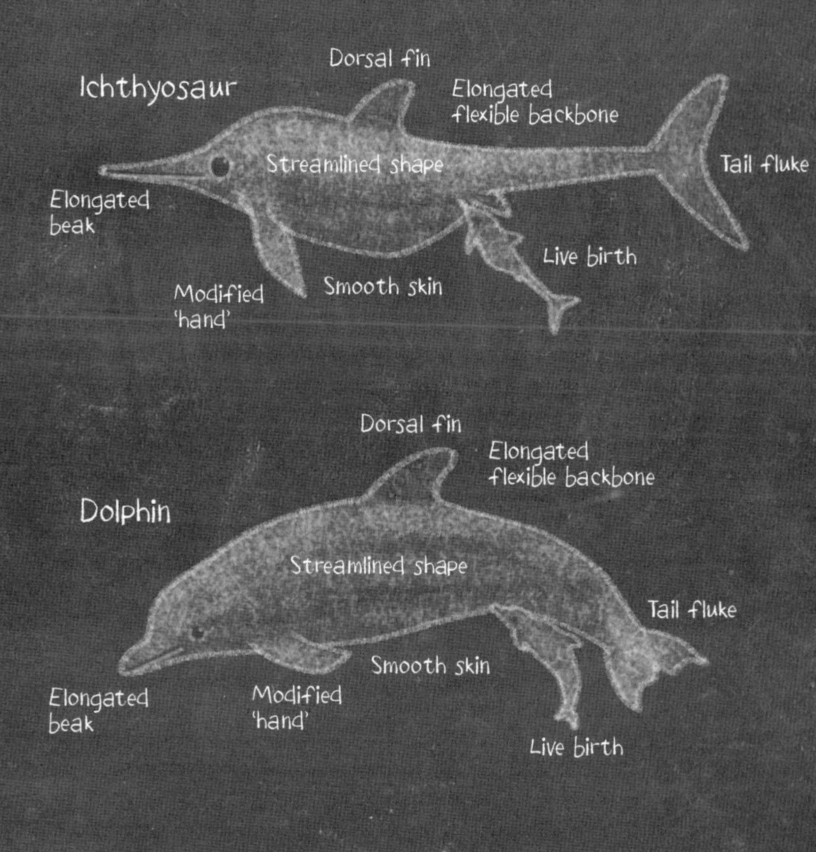
Ichthyosaur
Dorsal fin
Elongated flexible backbone
Streamlined shape
Tail fluke
Elongated beak
Live birth
Modified 'hand'
Smooth skin
Dolphin
Dorsal fin
Elongated flexible backbone
Streamlined shape
Tail fluke
Smooth skin
Elongated beak
Modified 'hand'
Live birth

Population genetics

The field of population genetics considers how the frequency of alleles in a gene pool can change. One of the biggest drivers of change is natural selection, but this is not the only thing that can alter the gene pool. Mutation is another factor. The rate at which viable mutations (ones that do not die out rapidly) appear is very slow, but over a long enough period they can be seen to produce regular changes in the gene pool. More rapid changes are introduced by phenomena called 'genetic drift' and 'gene flow'.

Genetic drift is caused by the element of chance. A freakish catastrophe may wipe out a significant proportion of a population and certain alleles may disappear along with it. A more mundane possibility is that alleles are simply not passed on, not because of selection, but merely through all the random aspects of the inheritance process. Gene flow, meanwhile, is the result of novel genes entering the gene pool with the arrival of individuals from another, hitherto isolated population of the same species.

Female choice

Natural selection is driven by the need for reproductive success, but males and females have different ways of achieving it. This come down to a difference between the sex cells known as anisogamy. Sperm contains only a half set of genes and they are easy to produce in copious amounts. A male's best option is to spread them as far as possible, playing the numbers to produce many offspring. The female's options are very different. Her eggs are primed with the energy needed to produce an embryo, and so are produced in much smaller numbers than sperm. After fertilization, the female must devote considerable resources to giving her offspring the best chances of survival – and she cannot rely on the male for help. Therefore, females make use of the shortage of eggs compared to the supply of sperm through the phenomenon of female choice: a female must choose, and choose carefully, how she wants to use her valuable reproductive resources. This creates a new element of competition among males that has had far-reaching effects on their evolution.

Two black grouse cocks compete for the best display position. Their mating success depends on being chosen by a female.

Sexual selection

In 1871, Charles Darwin published *The Descent of Man and Selection in Relation to Sex*, in which he expanded on his concept of 'sexual selection'. This form of selection, driven by mate choice (and largely female-led), does not necessarily lead to adaptations that aid survival. In fact, it can do quite the opposite.

Many of the impressive adornments seen in the animal kingdom, such as the antlers of a moose or the tail feathers of a peacock, are the result of this process. Darwin saw that sexual selection could outpace natural selection to create features that hindered survival. Taking antlers as an example, a female chooses a mate because he has large antlers. Any male offspring will grow large antlers as well, and any female offspring will chose mates with large antlers. This creates positive feedback that drives antlers to get bigger and bigger – far beyond their practical application as weapons. The result is that the sexes frequently evolve in different ways, creating marked differences known as 'sexual dichotomy'.

Red Queen Effect

Mate choice in animals frequently involves signals such as antlers, bright tails or some other adornment. Such signals are driven to extremes by sexual selection, but are nevertheless 'honest'. There is a high cost to developing large, symmetrical antlers, and that cost signals that the stag's genes as a whole are able to tackle the everyday requirements of survival and still have spare energy for growing large, often unhelpful antlers. A wonky-antlered rival's genes are less suited to survival.

But there is another factor at play that is keeping the antler signal honest. The population of deer (or any species) is constantly under attack from parasites and pathogens that are evolving unseen to get around an animal's defences. The deer evolve to counter these attacks – and those that succeed show it with their antlers. Although the species appears to remain unchanged, evolution is running all the time in the form of the 'Red Queen Effect', so-named for an *Alice in Wonderland* character who runs fast but always stays in the same place.

Human evolution

Darwin's theory of evolution met with many opponents, and the most controversial aspect of his thesis was that humans were produced by the same mechanism of change as all other life forms. This clash of ideas was epitomized in the 1860 debate between Darwin's ardent supporter Thomas Huxley and Samuel Wilberforce, Bishop of Oxford, when Wilberforce asked his opponent, 'Is it on your grandfather's or your grandmother's side that you claim descent from a monkey?'

Darwin had proposed that the anatomical similarities between humans and other primates, most notably the apes, showed that these species were our closest relatives. DNA evidence has since proved that humans share 98.8 per cent of our genes with chimpanzees and bonobos, while fossil evidence suggests that humans and chimps share a common ancestor that lived around 8 million years ago. Chimps remained as forest creatures, while humans evolved in a different direction as they became adapted to live on open grasslands.

Lucy

The earliest hominine (humanlike) fossil is *Sahelanthropus tchadensis*, a chimplike ape living in the forests of what is now Chad. Remains suggest it could stand on its back legs, but there is no evidence that this animal was a direct ancestor of modern humans. The earliest proven ancestor is 'Lucy' (*Australopithecus afarensis*), a bipedal ape specimen that lived in East Africa 3.2 million years ago (mya). Lucy was little more than a metre tall when she walked on two legs. However, her arms and fingers were considerably longer in proportion than ours, suggesting that Lucy and her fellow australopithecines ('southern apes') were able climbers, probably living in open savannah woodlands.

The lineage from Lucy to you and me is not clear but involves *Homo habilis* (*c.*2.5 mya – an omnivore that made simple stone cutting tools), and *Homo ergaster* (1.9 mya – thought to be the first species to control fire and spread beyond Africa). Several other hominin spread across Asia and Europe before *Homo sapiens* appeared about 150,000 years ago in Africa.

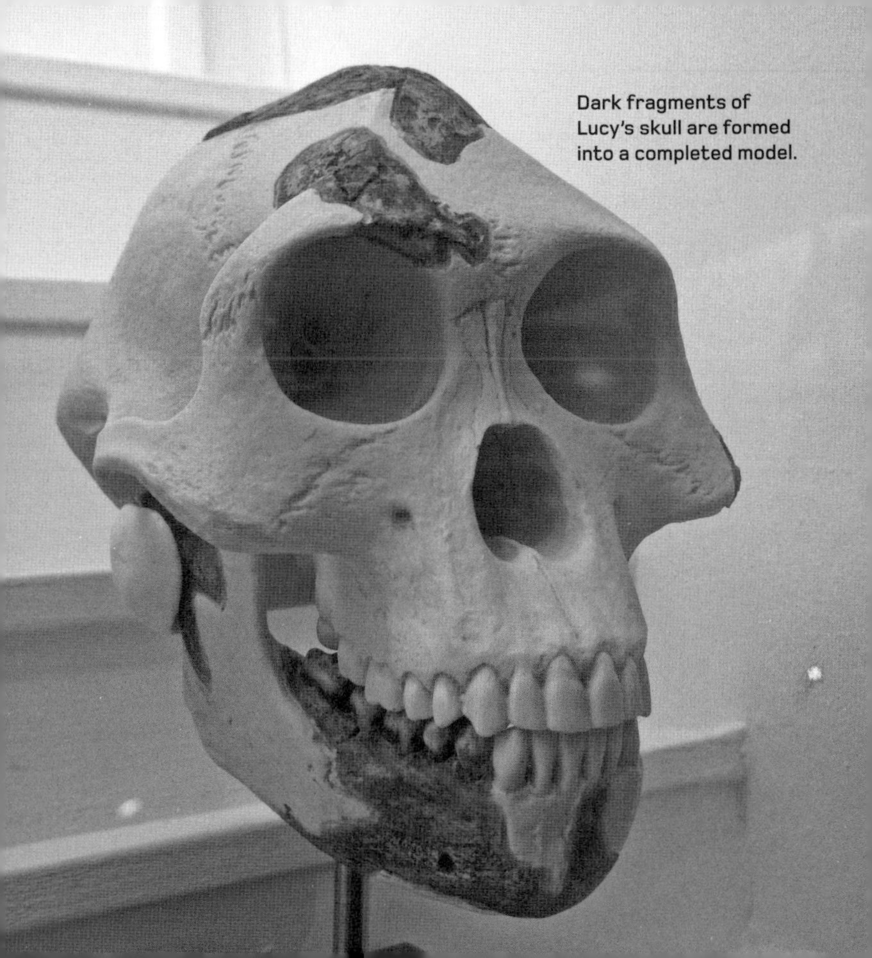

Dark fragments of Lucy's skull are formed into a completed model.

Classification

The modern science of biology and genetics arises from a growing understanding about the diversity of life on Earth. The first biologists were those people who sought to organize life into meaningful groups. This is the process of classification – or to give it its more scientific name, 'taxonomy'.

In the modern context, the work of taxonomists is often seen as the less exciting end of the biological sciences, a labyrinthine field associated with the minutiae of anatomy and lugubrious Latin names. So far taxonomists have classified around 1 million species, and the best estimates suggest this is barely 10 per cent of those living on our planet today. The modern field has changed beyond all recognition from the days when taxonomy was concerned with making drawings of specimens and giving them obscure names. Today, it is as much about tracing the ancestry of organisms throughout natural history as it is about parcelling up life into ever more groups.

Carolus Linnaeus

The founding figure of taxonomy is Carl von Linné (1707–78), a Swede better remembered by his Latinized name, Linnaeus. However, Linné was by no means the first to attempt to organize life according to a set of rules. Aristotle was one famous forebear though he made many blunders, for example in assuming whales were a kind of fish (the word 'dolphin' means 'womb fish' in Greek). Aristotle's mistake was to group organisms according to their habitat and lifestyle alone.

Linnaeus corrected the whale–fish error (eventually) in his *Systema Naturae*, a method of classification he refined through the mid-18th century. The Linnaean system was based on anatomical features, grouping organisms in a series of ranked groups according to the number of shared features. Linnaeus' finished list contained around 10,000 organisms, 60 per cent of which were plants (he was a keen gardener). Many aspects of the *Systema Naturae* persist in modern taxonomy although the classifications have been greatly extended and revised.

Linnaean naming

Linnaeus took on a Latinized version of his own name because Latin was the lingua franca of European scientists at the time. Every great work was written in Latin to overcome the language barriers and allow knowledge to disseminate better. So it is little wonder that Linnaeus also chose Latin and Greek as the languages for giving official names to the organisms classified in his *Systema Naturae*. The tradition has stuck to this day because it removes all ambiguities.

Crucially, Linnaeus opted for a binomial system, giving every organism two names. So the animal known in English as a lion is *Panthera leo* – only the first name is capitalized, and both are italicized. *Panthera* is the generic name, while *leo* is the specific one. The generic name refers to the genus to which the lion belongs, shared with similar animals, such as *Panthera tigris* (the tiger), *P. pardus* (the leopard) and the other 'big cats'. The specific name is there to delineate the lions from other big cats.

CLAVIS SYSTEMATIS SEXUALIS.

NUPTIÆ PLANTARUM.
Actus generationis incolarum Regni vegetabilis.
Florescentia.

PUBLICÆ.
Nuptiæ, omnibus manifestæ, aperte celebrantur.
Flores unicuique visibiles.

MONOCLINIA.
Mariti & uxores uno eodemque thalamo gaudent.
Flores omnes hermaphroditi sunt, & stamina cum pistillis in eodem flore.

DIFFINITAS.
Mariti inter se non cognati.
Stamina nulla sua parte connata inter se sunt.

INDIFFERENTISMUS.
Mariti nullam subordinationem inter se invicem servant.
Stamina nullam determinatam proportionem longitudinis inter se invicem habent.

1. MONANDRIA.
2. DIANDRIA.
3. TRIANDRIA.
4. TETRANDRIA.
5. PENTANDRIA.
6. HEXANDRIA.
7. HEPTANDRIA.
8. OCTANDRIA.
9. ENNEANDRIA.
10. DECANDRIA.
11. DODECANDRIA.
12. ICOSANDRIA.
13. POLYANDRIA.

SUBORDINATIO.
Mariti certi reliquis præferuntur.
Stamina duo semper reliquis breviora sunt.

14. DIDYNAMIA.
15. TETRADYNAMIA.

AFFINITAS.
Mariti propinqui & cognati sunt.
Stamina cohærent inter se invicem aliqua sua parte vel cum pistillo.

16. MONADELPHIA.
17. DIADELPHIA.
18. POLYADELPHIA.
19. SYNGENESIA.
20. GYNANDRIA.

DICLINIA (a δὶς bis & κλίνη thalamus s. duplex thalamus.)
Mariti & Feminæ distinctis thalamis gaudent.
Flores masculi & feminei in eadem specie.

21. MONOECIA.
22. DIOECIA.
23. POLYGAMIA.

CLANDESTINÆ.
Nuptiæ clam instituuntur.
Flores oculis nostris nudis vix conspiciuntur.

24. CRYPTOGAMIA.

Species

A species is the end result of the classification system. It denotes a group of organisms that share a large number of physical characteristics. One might think that members of a species always share more features with each other than they do with other members of their genus, and in the great majority of cases that is true – but not always. The crucial factor defining a species is that its members are all able to breed with each other, producing viable offspring that are fertile themselves.

This fact becomes important when considering 'cryptic species'. These are two populations of animals – often birds or bats – that are effectively indistinguishable by examining anatomy alone, but do not interbreed, and are therefore two distinct species despite looking more or less the same. Taxonomy can also classify organisms to a level lower than species. Many species are made up of subspecies – populations from different regions that may have significant anatomical differences – but are nevertheless able to breed with each other.

Taxa

Species and genus are examples of taxa (singular: taxon) – a word that derives from the Greek for 'arrangement'. The classification system puts every organism in a species, members of which share a small gene pool. Species are then placed within a series of increasingly larger taxa, which share larger gene pools. Each species belongs to a genus (plural: genera) and every genus has at least one species. The system continues by organizing genera into families. For example, *Panthera*, the big cat genus, belongs to Felidae, the cat family. Note that above the genus level, taxa are no longer required to be italicized.

The next taxon is the order. The Felidae belong to an order called Carnivora, alongside the Canidae (dog family), Ursidae (bear family) and other predatory animals. The Carnivora is one of the orders in the class Mammalia – the mammals. In turn, Mammalia is a member of the phylum Chordata, which includes other vertebrates such as reptiles, birds and fish. In botanical classification, the term phylum (plural: phyla) is generally replaced with the term 'division'.

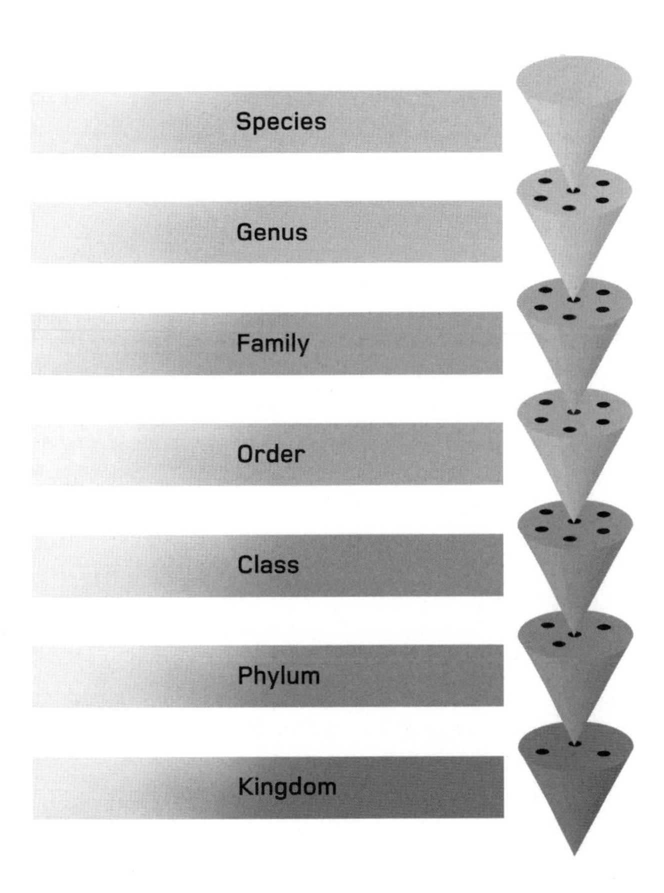
Species
Genus
Family
Order
Class
Phylum
Kingdom

Cladistics

At first glance, the classification system used today seems to follow the same methods developed by Linnaeus back in the 1750s. It retains the binomial names and the ranks of taxa. However, in the 20th century a new way of organizing life within the system began to dominate. This is cladistics, where species are classified not simply by the way they look and compare to each other, but how they are related by evolution. Every group of organisms that share a common ancestor is called a clade.

Cladistics requires that extinct species be included in the system along with the extant ones. When DNA is not available, taxonomists use statistical analysis of anatomy to find the most likely relationships between organisms, although the classifications produced this way are frequently challenged and changed. A good example of the implications of cladistics is the analysis of reptiles: natural history tells us that mammals and birds all evolved from the same ancestor as reptiles, and accordingly they all belong in the same clade: the Amniota.

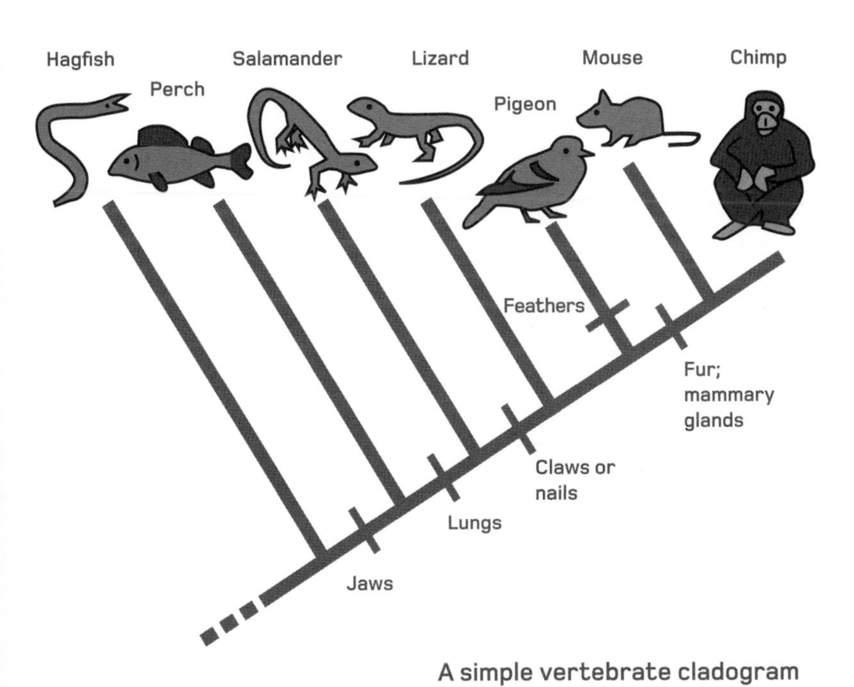

A simple vertebrate cladogram

Kingdoms and domains

One of the aims of classification is to create a big picture of life on Earth. Linnaeus did this using a final, top-ranked taxon called the kingdom. According to him, all life belonged to either the animal kingdom or the plant kingdom. The discovery of microscopic single-celled organisms created a problem: were they tiny animals or plants, or something else? Fungi were subsequently also split from plants, and the number of kingdoms went up. The simplest system used five: Animalia, Plantae, Fungi, Protista (amoebae, etc), and Bacteria. Then in 1977, DNA analysis showed that many cells that looked like bacteria were in fact a completely different set of organisms, now known as the Archaea.

Further analysis showed that the growing number of kingdoms could be grouped into three larger groups called 'domains'. Bacteria and Archaea occupy one domain each, while all other life lies within the Eukaryota – organisms with complex cells that use organelles (see page 60).

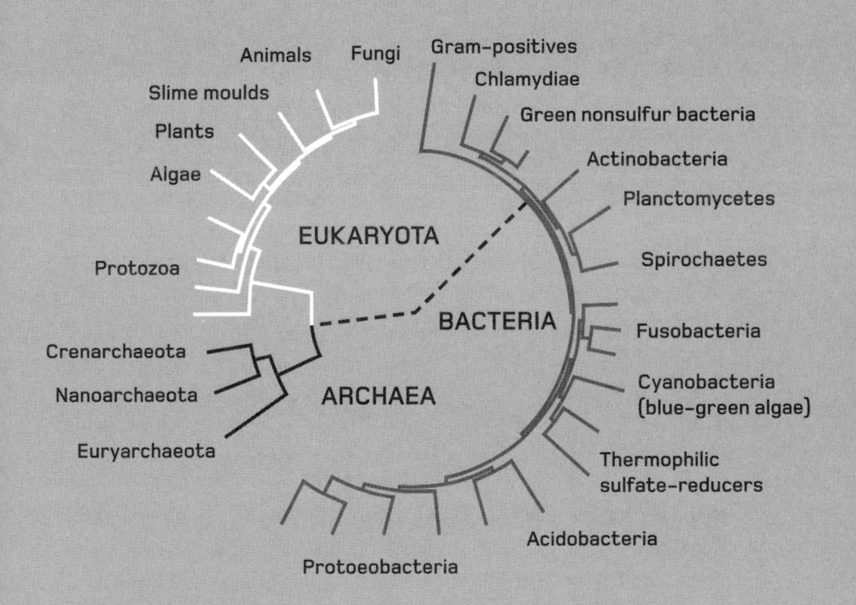

This diagram shows the proportion of life that lives in each domain. The kingdoms of plants, animals and fungi comprise just three branches.

Tree of Life

The only illustration in Darwin's *On the Origin of Species* is a branching diagram that shows how evolution by natural selection causes new species to radiate out from a common ancestor. Darwin visualized it more as a creeper or tangle of bushes growing on a steep bank, but the concept has subsequently become known as the 'Tree of Life'. It is still one of the best ways to visualize the great sweep of biodiversity that natural selection has created.

The trunk of the tree represents the primordial organism from which all life evolved. The trunk then branches into three domains, and each domain splits into kingdoms, phyla and so on. All the plants and animals fill just a third of the tree, and the mammals are represented as a mere sprig. Extant species form the tips of each branch, with the distances between them showing how closely related they are. Meanwhile, branches and twigs represent the intermediate, now-extinct forms they took as they evolved and split away from a common ancestor.

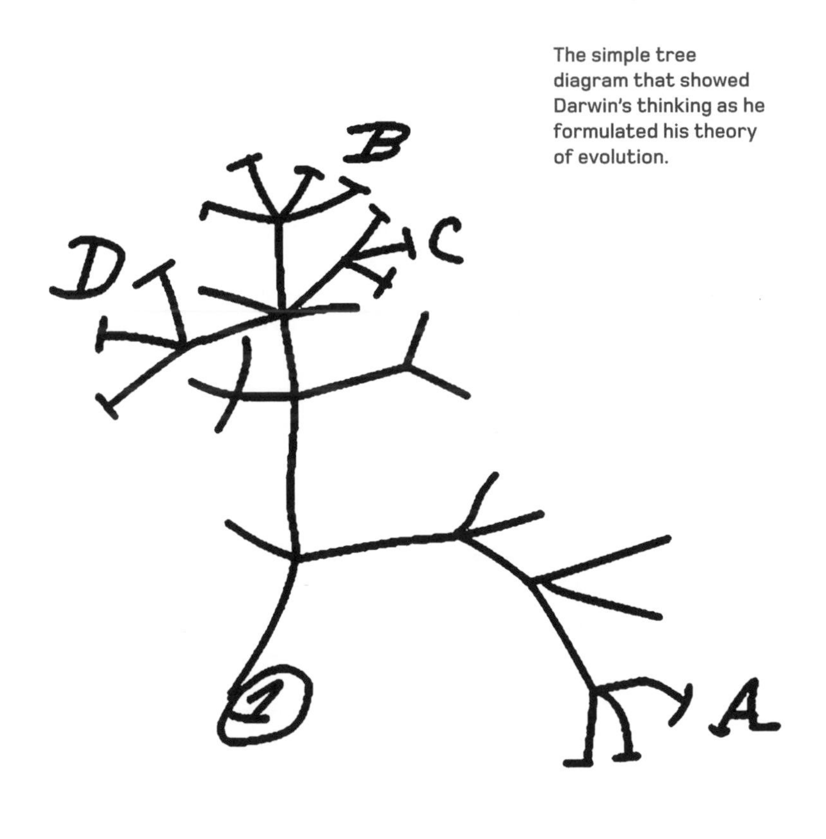

The simple tree diagram that showed Darwin's thinking as he formulated his theory of evolution.

Molecular clock

The fossil record of extinct organisms is far from complete, and the search continues to find specimens that might be common ancestors. However, genome analysis offers an alternative method for discovering when modern organisms shared an ancestor – the so-called 'molecular clock.'

Every species has a unique genome, at its most basic a long string of ACTG lettering (see page 92). The molecular clock system compares the differences in the letters between one species and the next. This is possible because while nuclear DNA is changed radically by recombination in every generation, the DNA in a cell's mitochondria is inherited directly from the organism's mother. Changes only occur as mutations, appearing at a more or less constant rate, like the ticking of a clock. Thus, there are fewer differences between closely related species than more distantly related ones, and the difference between mitochondrial genomes can be used to estimate when in the past the species diverged from their common ancestor.

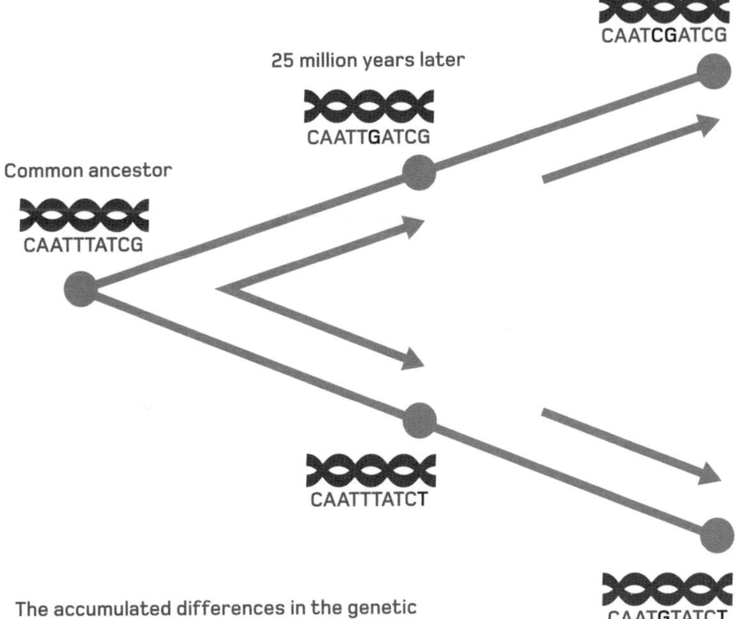

The accumulated differences in the genetic code of two organisms, arising from random mutations, provide an indication of the point of time in the past when the two species diverged.

Ecology

No organism stands alone; its genes and their resulting phenotype are all interacting with the environment. Ecology – meaning the 'study of the home' in Greek – is the scientific field that concerns the impact of those interactions. Ecologists are not simply biologists; they must also take into account aspects of geology, oceanography and climate science.

As if that were not enough, the real-life study of ecology is also exceedingly complex, if not impossible. Therefore, ecologists form models of wildlife communities. This could be for a specific habitat, or on a global scale – as espoused by the Gaia Hypothesis (which looks at the way that the Earth might function as a single self-regulating ecological system). The goal of the ecological models is to understand the factors that influence the success or failure of wildlife communities – and crucially to predict how they would be affected by human activities, such as habitat destruction and pollution.

Ecosystems

The term ecosystem is perhaps familiar but can be somewhat nebulous. Scientifically, an ecosystem is a way of describing a community of wildlife living in a particular habitat. However, in the real world, such ecosystems do not really exist in a meaningful sense, since there is no clear boundary between one community and the next. Nevertheless, the concept of an ecosystem is a good way of understanding the ecological factors at play in a certain habitat.

Every ecosystem is characterized by its ecological factors. These can be either biotic (living) or abiotic (non-living). Biotic factors are the interactions between different species present in the habitat, as they compete for food and resources, or prey and parasitize one another. Abiotic factors concern things such as soil chemistry, water supply and changing weather conditions. Ecologists model how an ecosystem would respond to changes in one of these factors – or the addition of a new one.

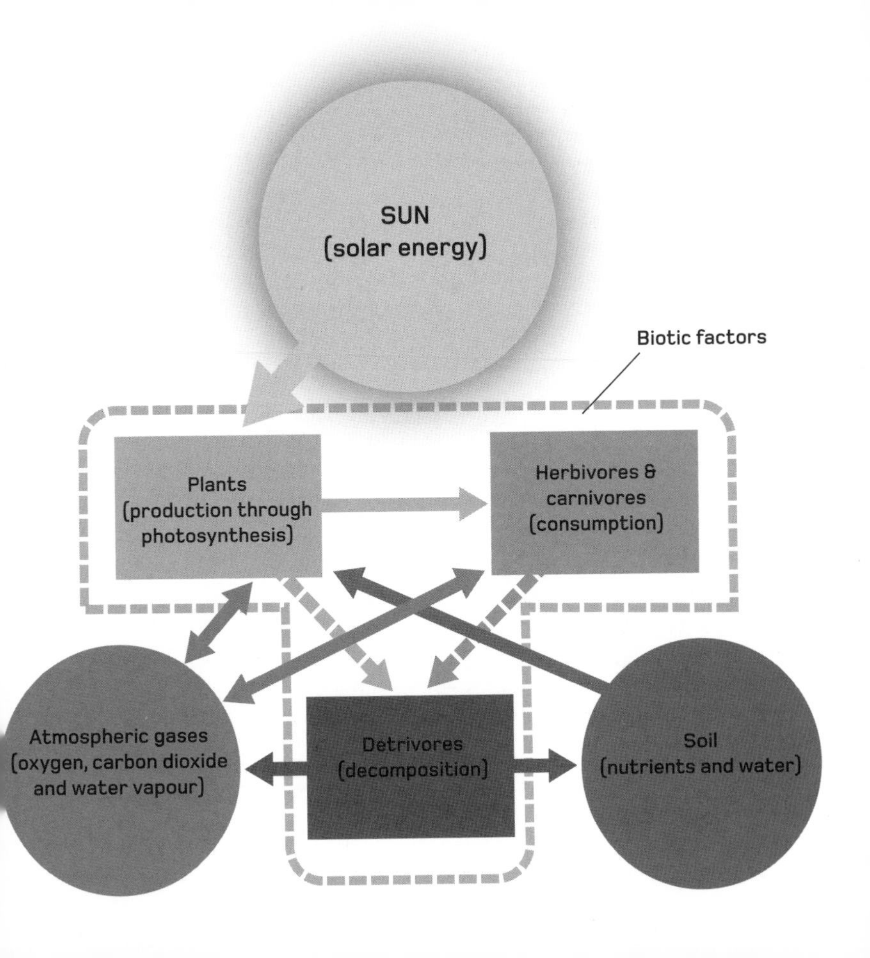

SUN
(solar energy)
Biotic factors
Plants
(production through
photosynthesis)
Herbivores &
carnivores
(consumption)
Atmospheric gases
(oxygen, carbon dioxide
and water vapour)
Detrivores
(decomposition)
Soil
(nutrients and water)

Ecological niche

Every species occupies an ecological niche within an ecosystem. A niche represents the opportunity for a life form to exploit the resources available in a habitat. As a result, natural selection adapts a species to survive in that niche by using a unique collection of anatomical features and behaviours. Speciation is driven by the presence of empty ecological niches, appearing either due to changes in abiotic factors (such as climate change), or when a population finds its way to a new habitat yet to be fully exploited by similar life.

The best known example of how organisms fill a niche are Darwin's finches. These birds (actually tanagers, not finches) live on the Galápagos Islands, and were seen by Darwin during his visit aboard the *Beagle*. The birds all descend from a common ancestor that arrived from South America, but have since evolved to occupy different niches in the islands' ecosystems. This is illustrated by bill shapes adapted to exploit a particular food supply, from insects to ripe seeds and fallen fruits.

1. Geospiza magnirostris.
3. Geospiza parvula.

2. Geospiza fortis.
4. Certhidea olivasea.

Habitats

Put simply, a habitat is the place where an organism lives. One could list dozens of distinct types of habitat and increase that list innumerably by making more specific descriptions of each one. Nevertheless, in general terms a habitat is an area where ecologists can construct a meaningful ecosystem. It could be a coral reef, a grassland or a tropical forest. There are common features within all these habitats irrespective of where they appear on Earth – and also commonalities among the organisms that comprise their wildlife.

Habitats are not constant. When a tree falls in a forest, for example, an empty gap is created. Organisms race to occupy this gap, with fast-growing plants arriving first, and then being gradually replaced by a succession of larger plants that are able to slowly but surely take control of the available resources. Eventually, the gap is filled completely and the habitat returns to its stable, or climax, state.

Biomes

The largest unit in ecology is the 'biome'. Each biome is a broad grouping of habitats that exist around the world. The number of possible biomes varies depending on scientific opinion, but this list is a good starting point: aquatic, forest, grassland and desert. Adding in the effect of different climates around the world, the list can be extended: deserts all have very low levels of liquid water, but can be split into hot deserts, semideserts and polar regions. Forests appear in areas of high rainfall and divide into tropical, temperate (home to deciduous trees) and boreal (conifer). Grasslands are places where the climate is too dry for trees to grow but not so dry that they are desert. They can be subdivided into three biomes: savannah, steppe or prairie, and tundra. Lastly, aquatic habitats can be divided between the saltwater marine biome and freshwater lakes and rivers. There are other ways of listing and defining biomes, but each results in a means to divide up the surface of the planet into large-scale regions for which the broad sweep of ecological factors is the same.

A map of the world's major biomes

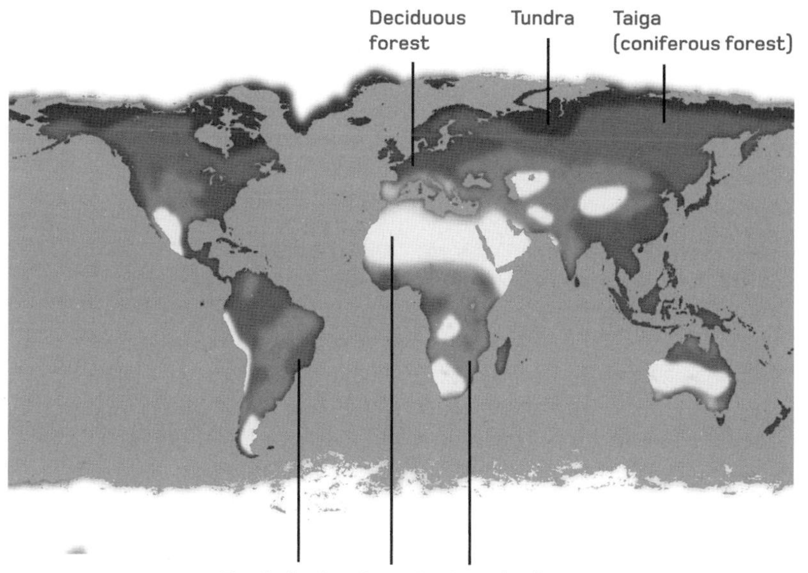

Food chains and webs

One of the most dominant factors within an ecosystem is what organisms eat. Relationships can be made into a food chain, or more pertinently, a food *web*, where many organisms are a possible meal for more than one other species. Food chains contain all the organisms in the ecosystem, and an almost universal feature is that the first point in the chain is a photosynthetic organism, such as a plant. These organisms are called primary producers because they collect energy from an abiotic source (sunlight) and make it available as a biotic resource. All other organisms in the chain are known as consumers. Herbivores, which eat only plant material, are called primary consumers. In turn they are eaten by secondary consumers and so on. Many secondary consumers are likely to be omnivores, meaning they eat both plant and animal foods. Tertiary consumers are almost certainly carnivores, restricted to a meat diet. Further up the chain we reach detritivores, such as dung beetles or fungi, which consume the waste and remains of other organisms.

African savannah food web

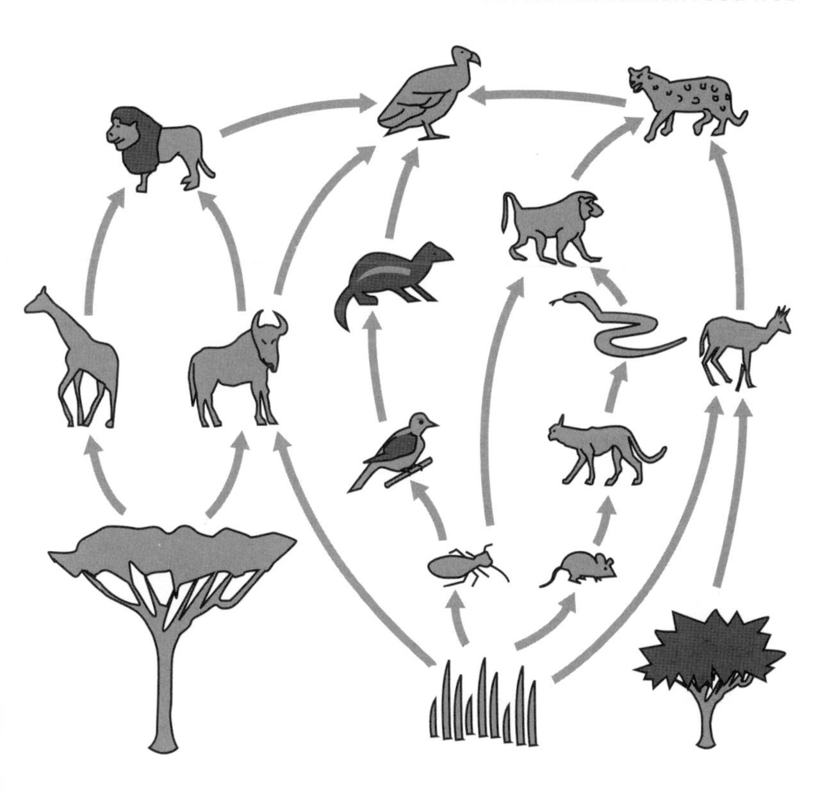

Trophic levels

A food chain illustrates routes taken by nutrients and energy through an ecosystem. The nutrients follow cycles, collected from the surroundings by plants, passed through the consumers and finally returned to the environment by detritivores. Energy does not work in the same way, but enters the system via primary producers and is then steadily lost as it moves along the chain. This is where the concept of trophic levels arises.

'Trophic' derives from the Greek word for 'feeder', and every step in a food chain moves up a trophic level. When all the trophic levels are presented according to the amount of energy – or more simply, by their biomass, or weight of living material – the food chain forms a pyramid. This is because only about 10 per cent of the energy from one level is passed to the next one up. As a result, the mass of plant material is far greater (in land ecosystems, at least) than the mass of animal material. This also explains why only a few large predators can survive in an ecosystem – hardly any energy makes it up to their niche.

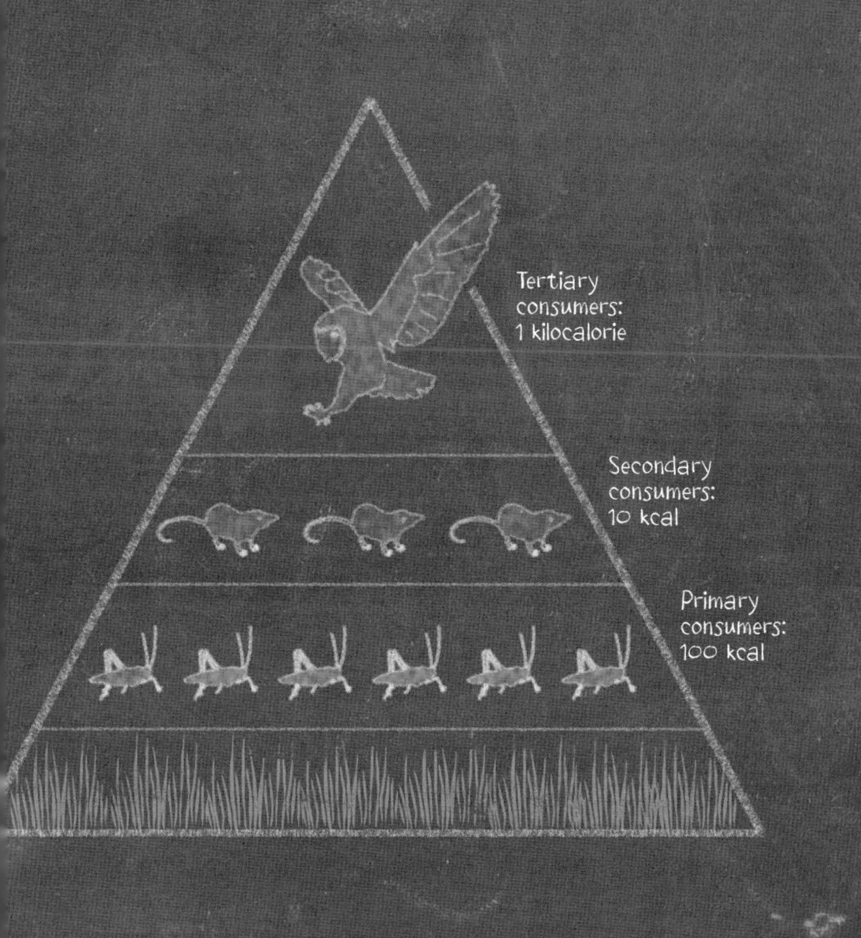

Tertiary consumers: 1 kilocalorie
Secondary consumers: 10 kcal
Primary consumers: 100 kcal

Autotrophs

There are many ways of being alive, and millions of distinct species that each do it in a unique way. However, they all split into two neat groups: autotrophs and heterotrophs. Primary producers belong in the first group – the term autotroph means 'self feeder', and refers to organisms that can harness a non-living sources of energy to power their bodies.

The most obvious examples are photosynthetic plants, seaweeds, and a whole host of microscopic organisms, including many bacteria. These are phototrophs, or 'light feeders'. However, some microbial autotrophs are chemotrophs, using chemicals as a supply of energy (see page 250). What an autotroph can do, and a heterotroph cannot, is 'fix carbon', taking inorganic forms of the element – chiefly carbon dioxide – and converting them into organic sugars. They do this using a chemical process called reduction – the exact opposite of the oxidation that releases metabolic energy during respiration.

Heterotrophs

The word 'heterotroph' means 'other feeder' - a reference to the way that heterotrophs are unable to fix carbon by themselves. Instead, they must get all their sugars and other raw materials by consuming the bodies of other organisms. All animals and fungi are heterotrophs, and many microorganisms live this way as well, including amoeba and protozoa. (*Euglena*, a kind of single-celled flagellate, is able to photosynthesize and consume at the same time.)

Heterotrophs rely entirely on autotrophs for their survival; even though a lion never eats any green vegetables it only survives by eating gazelles that have done. Consumers are not limited to predators and plant-eating prey. The simplest animals, the placozoa, are little more than mobile mats of cells, absorbing whatever organic particles touch the body. Meanwhile fungi (which include the largest organisms on Earth, sometimes spreading across 10 square kilometres of soil) exude digestive enzymes directly onto their foods and then absorb the results.

Chemotrophs

Until the late 1970s, it was assumed that all food chains began with phototrophic producers harvesting the energy from sunlight. Then, deep-sea submersibles discovered the hydrothermal vents known as 'black smokers'. These volcanic outlets release chemical-rich water on ocean floors far too deep for sunlight to reach. Yet despite the darkness, black smokers harbour an amazing ecosystem of giant worms, shellfish and crabs. The producers at the root of these food chains are prokaryotic (bacterial and archaean) chemotrophs.

Chemotrophs use the chemical energy of minerals in the vent water (or other volcanic sources, even inside rocks) to 'fix' carbon – some biologists believe they mght have been the first forms of life on Earth. The larger animals have evolved to live in harmony with these chemotrophs: some filter bacteria from the water, while the enormous tube worms host bacteria in their body tissue, providing safe harbour and a supply of minerals in return for sugars and other fixed carbon molecules.

Extremophiles

As their name suggests, extremophiles love extreme environments. The vast majority of life lives in a temperature range around 0–40°C (32–104°F), but the discovery of deep-sea hydrothermal vents revealed that bacteria and archaea could survive in superheated water above 100°C (212°F). They also thrive in hot volcanic springs at the surface, often creating a stunning rainbow of seemingly unnatural colours in the water.

There are extremophiles – almost always prokaryotes – that survive in regions that are very dry, super salty or acidic. There are even organisms called cryptoendoliths that live hidden inside rocks, occupying the tiny spaces between crystal grains. All of the above are certainly extreme to us, but in many ways these extreme habitats are more stable than our own, which is prone to all kinds of rapid and unpredictable changes. And when we look back into the conditions on the young Earth billions of years ago, today's extremes look rather normal.

The Grand Prismatic Spring in Yellowstone National Park in Wyoming, USA is a haven for extremophile bacteria.

Mimicry

Animals can be masters of disguise, using camouflage to blend in with their surroundings, using a body shape that looks like a leaf or a twig, or having a body pattern that makes them indistinguishable from a tree trunk or rock. But other animals hide in plain sight by mimicking the appearance of another member of their ecosystem – or even their smells, calls or behaviours.

There are two basic types of mimicry. The most common, called Batesian mimicry, involves a harmless mimic modelling itself on a dangerous neighbour. Thus, a hover fly has the stripes of a stinging wasp or bee, and many butterflies have dark eyespots on their hindwings, which resemble the face of a much bigger beast when opened up. The second type, Müllerian mimicry, is more nuanced, and involves species that do harm to attackers if eaten, such as poisonous butterflies. Predators learn to avoid toxic prey, and by evolving to look the same, both mimics are able to benefit from this lesson more effectively.

Coevolution

It is hard to imagine in this age of unrestrained habitat destruction and transformation, but some ecosystems have survived more or less unchanged for many millions of years. In all that time species have evolved in concert with each other through a phenomenon called coevolution, where a change in one species triggers a change in a second in reaction to the first. This creates a complex of interlacing adaptations that allow an ecosystem to support a large capacity for life. However, that strength is also a weakness because sudden changes from outside the system – typically human activities – are able to easily disrupt the fine balance between life forms.

Classic examples of coevolution are the arms race between predators and prey, and the adaptations of flowering plants and insects or birds that work in symbiosis. By evolving a co-dependent relationship, the plants increase the chances of their pollen being delivered to others of their species, while the animals find themselves with a reliable food supply.

Animal relationships

Many animals are solitary – they just want to be alone. Primitive asexual creatures, such as a budding hydra, need no one else to make a success of their lives. However, most animals engage in some kind of relationship with another member of their species that is of mutual benefit to both parties. This may be as simple as pairing up with a mate, or it may be part of a more complex society where members of the same species cooperate to a lesser or greater extent.

For example, a piece of coral is actually a colony of thousands of individual animals called polyps. The polyps grow side by side, but feed and reproduce as individuals. However, at the microscopic level, members of the colony also work together to fend off encroachment from neighbouring corals. To understand animal societies and other animal relationships we have to weigh up their various benefits and disadvantages. For example, living in a group increases competition for food and mates but it also boosts safety and defence.

Symbiosis

A symbiotic relationship is one in which members of two different species have evolved to live closely together for mutual gain. Flowering plants and honeybees exist in a symbiosis, for example: neither could survive without the other, and the partnership has come about by an extreme form of coevolution. Some symbioses are an even closer union. Giant clams and corals harbour photosynthetic bacteria called Zooxanthellae in their tissue. These so-called endosymbionts provide their hosts with sugar in return for a safe and stable place to live.

There are two modes of symbiosis. The examples above are mutualistic, with both species gaining from the relationship. But there are some symbiotic pairings where only one of the partners gains. The other one neither gains nor loses. This situation, known as commensalism, is less common than mutualism, but examples include cattle egrets (opposite) that follow herds of cattle or other large herbivores to prey on the ground insects that are disturbed by the herd's hooves.

Parasites

Coevolution can also create relationships between two unrelated species that benefit one partner and damage the other. This is parasitism – the beneficiary in the partnership is the parasite, and the loser is the host. Parasites can live inside or outside the host. Fleas (opposite) are ectoparasites, living on the skins of their furry hosts, while a tapeworm is an endoparasite, living in the gut of its host. Other parasitic worms enter the body proper and set up home in the blood and organs.

The life cycles of parasites are often convoluted, with parasites transitioning through several hosts, or 'vectors'. For example, mosquitoes are the vectors for the malaria parasite, while river snails carry schistosomes, tiny worms that cause the tropical disease bilharzia. The most successful parasites do little harm to their hosts – killing a host means they must find a new place to live. However, some organisms called parasitoids do kill their hosts – eventually. Such animals (often tiny wasps) lays eggs on or inside the host, and the young slowly eat it alive.

Predators and prey

A predator is an organism that kills and eats other organisms. There are a few examples of plants and fungi as predators (the Venus fly trap is famous for trapping and digesting insects, while some fungi snare microscopic worms in the soil). However, the great majority of predators are animals, and the organisms they target are their prey.

The common usage of these terms suggests that predators are fierce meat-eaters, frequently large and powerful beasts that are able to overpower their weaker prey. However, an assassin bug stalking a cricket or a ladybird larva grabbing an aphid has all the drama and violence of predator–prey relationships among larger animals. And there is also a dynamic relationship between the populations of any predator and their prey. When the prey population is large, predator numbers grow thanks to the surfeit of food. But with so many predators around, prey numbers begin to drop – and predators starve. With fewer predators, the prey population rises once more, and the cycle begins again.

Predator–Prey cycles

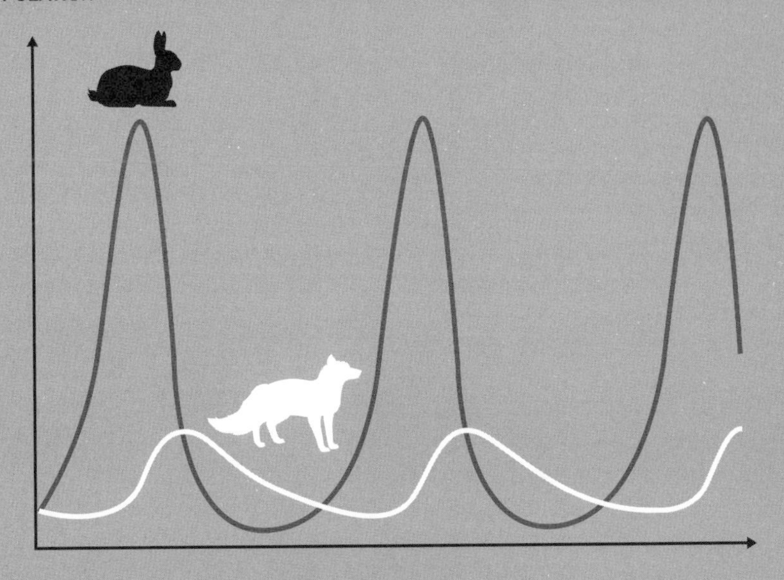

Herding

Herds go by many names: flocks, shoals, even gaggles. All these animal groupings are loose, leaderless affiliations. The members are essentially living solitary lives, acting to maximize their own success, but this goal is best served by staying close to others of their species. It can look like members of a herd are all working together but this is an effect of them all behaving in the same way.

The most obvious reason to live in a herd is safety in numbers. Predators will attack individuals on the edge of the herd, and so those animals that find themselves on the periphery are always moving towards the centre, keeping the herd together. Each herd member remains on the lookout for danger, and moves to safety when it is alarmed. As a result, the whole herd is soon alerted to the threat and moves away *en masse*. Herding behaviour suits animals that live in habitats where food is widely distributed, such as grazers on a grassland. Animals that exploit small concentrated supplies of food, such as fruit-eaters, would be at a disadvantage if they moved in such large numbers.

Polygyny

Polygyny is a mating system where one male mates with several females. By contrast the female mates with a single male, perhaps because she becomes unreceptive to further courtships, but more often because her male mate keeps guard over her and his other mates. This creates a 'harem', where a single male controls the reproduction of a group of females and keeps other males away.

Many group-living animals, such as deer, hippos and cattle, use a polygynous mating system, as do many fish. A common factor is that these animals live in habitats with distributed food supplies. This makes it impossible for one animal to take control of a meaningful feeding territory. There is no point in a stag defending a parcel of land to control food resources, since his rivals will find plenty of food elsewhere. So to maximize success, a stag takes control of reproductive resources instead. This strategy leads to distinct sexual differences as males become fighting machines able to defend a harem.

Polyandry

Polyandry is the opposite of polygyny. A single female has several male mates, who are exclusively hers – they do not mate with other females during that breeding season. Polyandry is rarer in nature than polygyny, but it does exist. An extreme form is seen in anglerfish: the seafloor predators that lure prey with a glowing lantern are all female, since males never develop beyond their juvenile form. Instead, they bite into the body of a larger female, eventually becoming linked to her blood supply. Each adult female will have several such mates embedded in her skin, shedding sperm whenever she lays eggs. Other examples of polyandry are more conventional: spiders, reptiles and several birds, including the emu, all follow this strategy. Some females store sperm in order to produce a single clutch of young with multiple fathers. This means that male mates are all willing to share paternity and help the female look after all the young, since they do not know which ones may be theirs. In polygyny, males devote all their time and energy to mating. In polyandry, the female must shoulder a similar burden but also produce the young as well.

Promiscuity

If we strip away the moral connotations of this word when applied to human behaviours, we find that many animal societies employ 'promiscuous mating' – including many of our closest relatives. Simply put, a promiscuous mating system is one in which both males and females mate with multiple partners – and do not form pair bonds.

This kind of sexual strategy is most common in so-called fusion–fission societies, as seen in many monkey and ape species, and in dolphins. Members of such societies generally cooperate to find food supplies, defend against dangers, and protect the young, irrespective of parentage. There is frequent mixing between groups, where two or more groups will meet and merge for a while and then split again into smaller units with a different set of members from the original ones. Such groups may have a leader – or a dominant leadership group of older individuals – but there are no barriers to members moving between groups.

Mixed sexual strategy

The way animals breed is not always set in stone. Some species vary their mating system – or sexual strategy – to suit ecological conditions. Lions are a good example, famous for being the only cats that live in social groups. The most familiar grouping is the pride, in which one male controls a harem. This system works on the grasslands of Africa where lions must cooperate to catch fleet-footed herbivores. But in locations where food is easier to come by, lions form monogamous pairs, with one male and one female working together to raise their cubs. When lions lived in Europe, for example, they adopted this strategy.

Monogamy is the *de facto* system for many other animals. Forming long-lasting pairs reduces the need for males to expend energy competing for mates, allowing them to devote more effort to ensuring their offspring survive. Nevertheless, monogamous animals may mate elsewhere when they can – a female cheater gains by adding new genes to her offspring, while a male gains through another male raising his young.

Sexuality

Human sexuality has a social component beyond the realms of biology, bound up in ideas of sex (male or female), gender identity (masculine or feminine), and taboos and gender roles that vary from culture to culture. Is homosexuality caused by inheriting a gene? There is some evidence that both of a pair of identical twins being gay is a more likely outcome than just one, but this is thought to be an epigenetic, rather than genetic, effect (see page 354) – one proposed cause points to hormone levels in the uterus during pregnancy.

In human culture there remains a strong link between sexuality and personal identity, with individuals often required to declare as heterosexual, homosexual or bisexual. Do such distinctions in the natural world beyond humans? In a biological context sexuality can be treated simply as a set of behaviours. Homosexual behaviour is sometimes seen in many different animals, but is seldom the dominant mode during courtships: in species such as the bonobo, it provides a means of mediating complex social relationships.

Sex ratios

For the majority of species, the ratio of male to female is roughly 1:1. Individual mothers may produce families that are all female or all male, but the chance of them producing a male or female offspring is always 50/50 each time. The reason for this equality is summed up in Fisher's Principle, attributed to Ronald Fisher in 1930. If females outnumbered males, there would be an obvious advantage in producing males, since with a larger number of mates they would be able to sire more offspring. These offspring would also have a tendency to produce male offspring – pushing the sex ratio back towards 1:1. If it overshoots, a similar mechanism for the female sex would act to bring it back to a stable equilibrium. However, there are situations where the sex ratio remains skewed. For example, fig wasps spend their larval stage inside figs. The adults that emerge are mostly female, but there are a few males that mate with their sisters before they fly off to lay eggs in the next fig, thus removing the need for a balance of the sexes and maximizing reproductive output.

r/K selection

The rather obtuse phrase 'r/K selection' refers to the two main strategies for producing offspring. An r-selective strategy focuses an animal's resources on the rate of reproduction (the r stands for rate) while a K-selective one focuses on maintaining the animal population at full capacity (the K stands for *Kapazitätsgrenze*, or 'capacity limit'). A pinnacle of the r-selective species is the oceanic sunfish, the largest bony fish on Earth: a female produces 300 million eggs every year, by far the largest number of any vertebrate. Just a handful of these eggs will reach adulthood, but the fish is playing the numbers. If it can produce a few tens of millions of young more than its neighbour, it should have more success – and its babies will be able to take advantage of whatever openings in the ecosystem may arise. At the other extreme is the archetypal K-selective species, the orangutan. A baby stays with its mother for seven years, learning all it needs to survive in the forest. Only then will the mother have another child, thus limiting her lifetime fertility to an average of just two young.

The oceanic sunfish and the orangutan use very different strategies to achieve reproductive success.

Sex determination

The mechanisms that determine an animal's sex are not universal. Most higher animals, such as birds and mammals, use a genetic system, but other methods are at play elsewhere in the animal kingdom. Mammals – including humans – use the sex chromosomes X and Y. A female has XX, while a male has XY. Females are homogametic, meaning a female gamete always carries an X chromosome. The male is heterogametic in that half his sperm carry an X and the other half a Y. Birds also use a genetic system, involving the ZW genes. In this case, however, the males are homogametic, with ZZ, while females are heterogametic with ZW. Many insects use a similar system where females have two sex chromosomes, and males have just one. The sex of crocodiles, turtles and some other reptiles depends on the temperature of the nest. In turtles, eggs that are at a lower temperature tend to be male, with the rest being female. In crocodilians, eggs that are in the mid-range of temperatures are male, while hot and cold ones become female. This system is prone to wide fluctuations year on year.

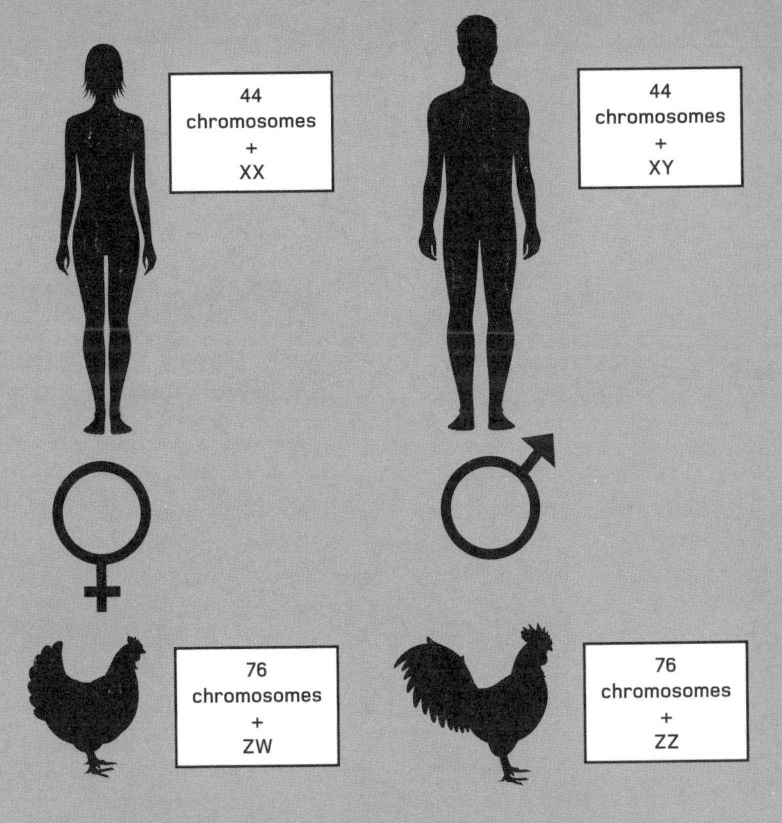

Chromosomes and sex determination

Eusociality

Ants, termites and honeybees are all examples of eusocial animals. They live in colonies where the offspring of a single queen act as workers that build a nest, collect food and raise more of their mother's offspring. This kind of social unit thrives in arid areas where individuals would struggle without cooperation. A termite queen (opposite) is a giant egg-producer, many times bigger than the workers, with a king who lives alongside her: both male and female offspring that are kept infertile by pheromones she releases. Occasionally, fertile winged offspring are sent out to reproduce and start a new colony. In contrast ant, bee and wasp workers are all female. They and their mother, the queen, are diploid, while unfertilized haploid eggs develop into males, or drones, that leave the nest along with virgin queens to mate and start new colonies. The genetics are complex, but the system means that the sisterhood of ant workers are more closely related than normal sisters. This in turn ensures the workers will devote themselves to helping their mother to produce more siblings.

Sperm competition

Competition for mates – generally between males – can be fierce and often deadly. However, once mating is over the competition does not end. Sperm are the product of meiosis and as such they have a different genetic makeup to their creator, and different genes from each other. As a result, each sperm is in direct competition with its neighbours. A leading hypothesis as to why crossing over evolved (see page 134) is to reduce the genetic differences between sex cells and so lessen this competitive streak. If it was left unfettered, natural selection would make sperm literally attack each other. Instead, evolution has resulted in numerous adaptations where the sperm from different males compete. The simplest one is mate guarding. A male harlequin toad stays attached to his mate for 19 days to stop rivals mating. Some males secrete a plug that blocks up their mate's genitals. Meanwhile, in promiscuous species a male's penis may scrape out earlier sperm deposits before leaving its own, after which it all comes down to a trial of speed and stamina as sperm race to the eggs.

Animal culture

One advantages of a K-selective strategy, where parents invest time and energy in raising a few young rather than simply producing offspring in vast quantities, is the ability to teach the young. Many behaviours, from hunting techniques to social interactions, are learned in childhood and passed down from generation to generation. The learned aspect of an animal's behaviour has a cultural dimension because groups of the same species living in different parts of the world behave in different ways. A good example is killer whales: these 'wolves of the sea' hunt in packs, or pods, in all corners of the ocean. Yet each pod has a hunting style that suits where they live. Some target shoals of fish, others stalk whales, while others snatch seals – all using learned and well-practised cooperative hunting techniques. A fish-eating killer whale moving to a whale-catching group would struggle to fit in with the culture. Animal culture evolves and radiates in a way that mirrors natural selection, with novel behaviours taking root in one group before moving to another.

Neo-Darwinism

The modern understanding of evolution by natural selection is known as neo-Darwinism. It does not contradict the great Darwin's original theory, but instead merges it with a more recent understanding of inheritance and genetics.

Neo-Darwinism came to the fore in the 1960s through the work of John Maynard Smith and William Hamilton (later popularized by Richard Dawkins, pictured). Its most obvious contribution was the correction of a misunderstanding that had grown up around Darwin's original theory. As the theory became widely accepted in the early 20th century, it was assumed that natural selection acted at the species level, with adaptations made for 'the good of the species'. How natural selection did this was not clear, and the unit of selection was corrected to the individual body, which was more in keeping with Darwin's original idea. Today, evolutionary biology sees a body as a machine produced by genes to ensure their survival. Thus, natural selection is really only at work at the level of the gene.

Altruism

There are many reasons why humans are altruistic, or selfless, to one another. It might be due to a moral code, to boost personal virtue or to maximize benefit for the majority despite a personal cost. However, other animals also behave altruistically. For example, a worker bee will sting an attacker and condemn itself to death in the process; a meerkat sentinel (opposite) will bark a warning to its mob when it sees a predator, attracting attention to itself, while giving the others a chance to escape; and a monarch butterfly will risk being eaten by a bird – and make its assailant sick from the toxins in its body – in order to teach that bird to stay clear of related butterflies in future. How does this behaviour match up with the brutal struggle for survival at any cost that underlies natural selection? The only reason required is that parents and offspring share genes, so do other members of a family or social group, and indeed all members of a species to a lesser extent. Animal altruism may hinder, or even kill, an individual, but it also ensures the survival of many more copies of the genes it shares with others.

Relatedness

Also called the 'coefficient of relationship', relatedness is a way of putting a numerical value on how closely related individual organisms are in terms of the genes they share. An organism that reproduces asexually will share 100 per cent of its genes with its offspring, and so they have a relatedness of 1. Sexually reproducing organisms pass on half of their genes to their offspring, and although they may not be expressed in the phenotype, they are still present in the genotype, giving parent and child a relatedness of 0.5. The same is true of siblings; they have a 50 per cent chance of inheriting the same genes from their parents and so they have a relatedness of 0.5. Half-siblings and grandparents have a relatedness of 0.25, while first cousins are 0.125 related. Second cousins share about 3 per cent of your genes, which in the human gene pool makes them as closely related as a complete stranger. Relatedness has its place in human laws to avoid inbreeding (0.125 is usually the limit) but it also helps to understand behaviours in the wider animal world.

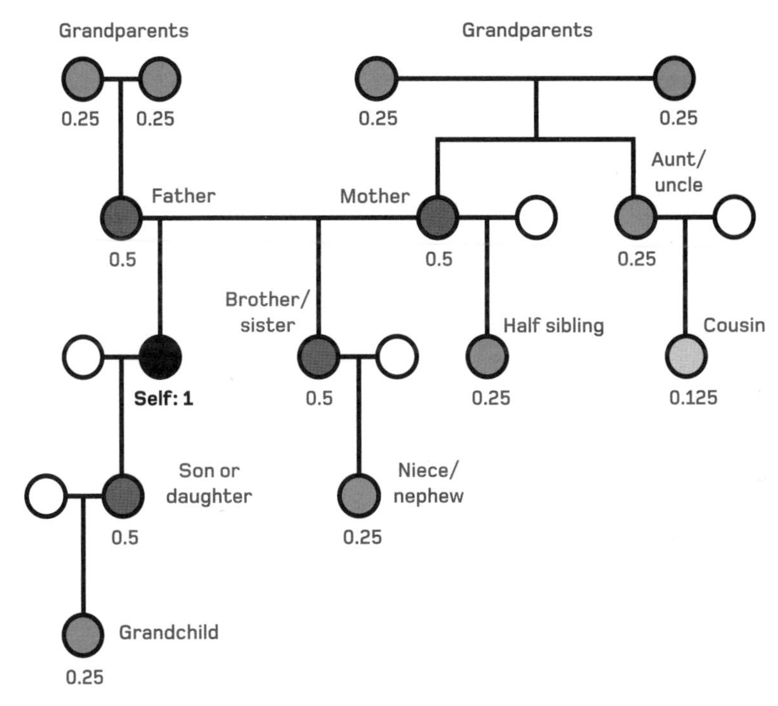

Degrees of relatedness

Selfish genes

Made famous by Richard Dawkins' 1976 book of the same name, the 'selfish gene' concept is frequently misinterpreted. It is *not* a declaration that our genetics determines our personality, giving us a licence to behave as selfishly as we like because it is the 'natural' thing to do. Nor is it saying that genes are constantly calculating – by some supernatural intelligence – the best outcome for themselves and acting accordingly.

Instead, the phrase 'selfish gene' was intended to sum up the central tenets of neo-Darwinism: that natural selection works at the level of genes; that the drive for survival is the result of a gene's drive to be replicated in ever greater numbers; and that all animal behaviours, even seemingly altruistic ones, can be explained in these terms. Genes do not exist in order to produce the phenotype (the organism's body and behaviours). In fact, the reverse is true: the phenotype is a means by which the genes can ensure their survival and maximize their abilities to replicate themselves.

Watching over her pups, this mother capybara (one of the world's largest rodents) is also protecting her genes.

Green-beard effect

Altruistic behaviour in animals is the result of selfish genes ensuring their survival. Parents protect their offspring from attack, and brothers and sisters help each other raise young. This is called 'kin selection', where an individual works to boost the reproductive success of their kin, and so replicate their own genes by proxy. But what if unrelated, non-kin animals could see that they shared at least one gene? Would they then help each other out? This is the question considered by a thought experiment called the 'green-beard effect'. Imagine a gene that gives the carrier a green beard. The gene also makes its carrier recognize other individuals with the green-beard gene and behave altruistically towards them. Such a system would be very advantageous to the gene, but does it exist in nature? There are a few possible examples, mostly in microorganisms, but the system is rare because it is prone to cheating: a mutant green-beard allele that precludes altruistic behaviour would get all the benefits with none of the costs, and rapidly replace the original version of the gene.

Taking risks to aid people that look like you is a flawed genetic strategy – only cheaters benefit.

Hawk–Dove

Another modernizing aspect of neo-Darwinism is its use of game theory in understanding the evolution of animal behaviours, especially those used in conflict resolution. Game theory is an arm of probability, the mathematics of chance. It was developed in the 1940s to help predict human behaviours in economic and military scenarios, but can be used in a simpler form to show how species evolve stable behavioural strategies.

The Hawk-Dove 'game' considers the benefits of being belligerent (a hawk) or a pacifist (dove). It pits two individuals against each other in a conflict over food or a mate. If a hawk meets a dove, he always wins the prize. If a hawk meets another hawk, he has a 50 per cent chance of winning (or losing). If a dove meets a hawk he never fights, wins nothing but loses nothing either. If two doves meet, they share the resource and get half each. The chances of these four scenarios occurring depends on the frequency of hawks and doves in the population. A hawk among doves does well, but so too does a dove among hawks.

Hawk–Dove model

Determinism

The emphasis that neo-Darwinism places on the power of the gene to influence evolution has led to a belief that genes are the single factor in the development of an organism's body and its behaviours. People think that having a certain gene will inexorably lead them to have a certain trait, saying things like 'it's in my genes' to explain away their behaviours and other features. This is a mistaken belief known as 'genetic determinism'.

Classical Darwinism refers to the interplay between an organism's body and behaviour and the environment. Darwin knew nothing of Mendelian genetics, and when the two ideas were fused, a shorthand emerged that seemed to suggest that an organism's genes were the *de facto* instructions for building a body. In fact, it was always understood that every gene interacted with the environment to develop a unique body. The question was how much of the end result is nature (the gene) and how much is nurture (the environment) – see page 346.

German biologist August Weismann (1834–1914) was the first great advocate of genetic determinism, and is often seen as a pioneer of the 'selfish gene' theory.

Behaviourism

The field of psychology known as behaviourism seeks to understand the way animals, including humans, learn, and aims to understand the motivations of an animal through its observable actions. One technique used to investigate learning is 'operant conditioning', in which a test animal is taught to perform tasks through a system of reward and punishment. This technique's most vocal exponent was American psychologist B.F. Skinner, who designed a chamber that could be used to house his test subjects. Skinner mostly used pigeons, which were rewarded when they performed correctly, reinforcing that behaviour. Skinner taught pigeons to perform complex sequences of tasks just as well as any more 'intelligent' test subject. His conclusion was that learning was a purely physical process that did not require any mental component – even in humans! This radical notion went unchallenged for 20 years until the late 1960s, when the first physical trace of a mental memory was isolated in the brain, proving a link between physical and mental phenomena.

Skinner box for a pigeon

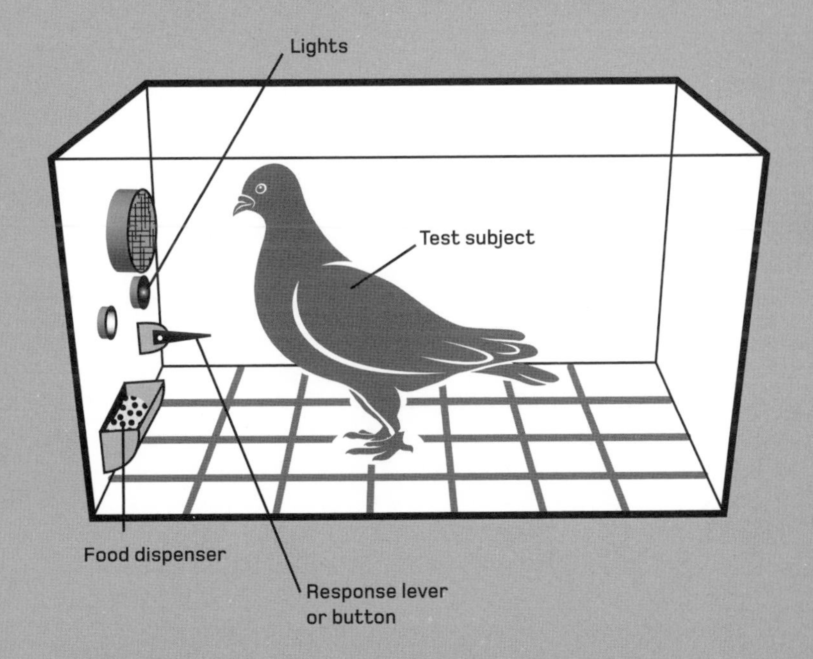

Origins of Life

The theory of evolution by natural selection has largely been accepted as the mechanism by which organisms can change over many generations. Central to Darwin's thesis was that all life evolved from a single primordial organism, but the characteristics of this common ancestor are unknown. It was certainly microbial, since the earliest parts of the fossil record (from around 3.5 billion years ago) contain only prokaryotic organisms such as bacteria and archaea. How these life forms – which are simple compared to us, but very complex compared to inorganic structures – could have arisen from non-living material is an enduring mystery, although there are many theories.

Darwin himself saw the origin of species and the appearance of life as two separate problems, and imagined 'in some warm little pond with all sorts of ammonia and phosphoric salts – light, heat, electricity etc., present, that a protein compound was chemically formed ready to undergo still more complex changes.'

Fossils known as stromatolites preserve many-layered colonies of simple microorganisms from up to 3.5 billion years ago.

Primordial Soup theory

The most famous theory concerning the origin of life reflects Darwin's own suggestion that it emerged from some 'warm pond' in the distant past. The 'warm ponds' in this case were the first permanent oceans that filled Earth's basins about 3.8 billion years ago. The theory was given its most significant boost by the 1952 Miller–Urey experiment, carried out by chemist Stanley Miller and astrophysicist Harold Urey at the University of Chicago. They assembled an apparatus named the Lollipop for the disc shape of its central reaction chamber. This chamber was seeded with water and chemicals found in volcanic emissions, such as nitrogen, carbon dioxide and sulphides. The mixture was stirred, boiled, condensed and electrified in a constant loop. Within a day it turned pink, and after a week the researchers found it contained many complex chemicals – cyanide, ammonia and even a simple amino acid. If the experiment was run on a larger and longer scale, the researchers reasoned, it would eventually produce all the chemicals of life.

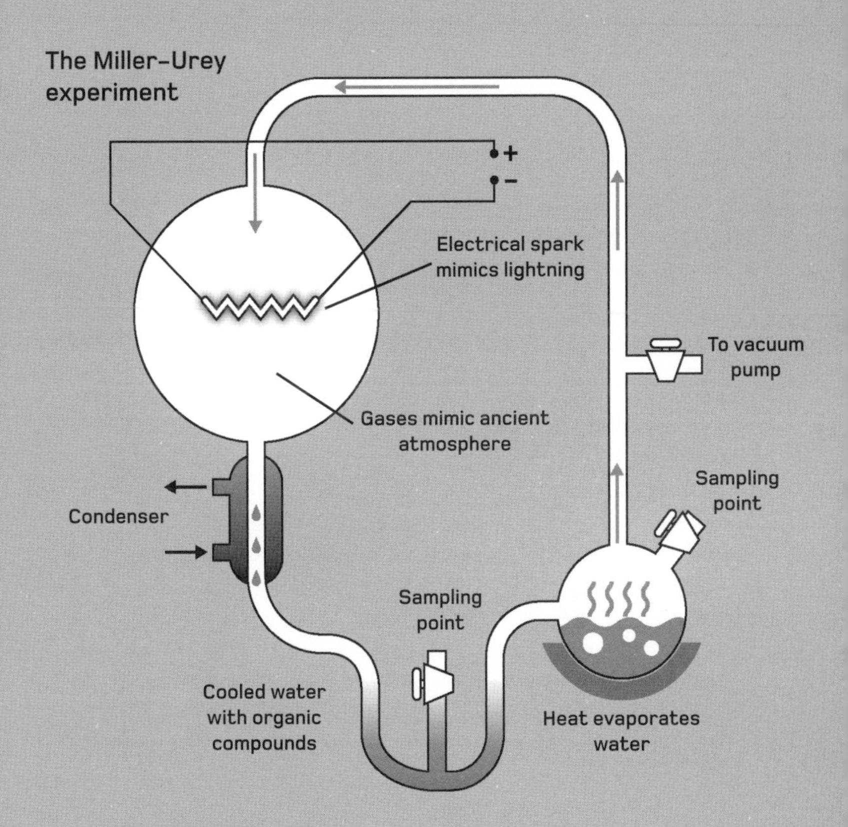
The Miller-Urey experiment
+
−
Electrical spark mimics lightning
To vacuum pump
Gases mimic ancient atmosphere
Sampling point
Condenser
Sampling point
Cooled water with organic compounds
Heat evaporates water

Panspermia

The first step in the appearance of life must have been a non-living chemical process capable of building the biological chemicals used by life. However, the 'panspermia' theory neatly sidesteps this issue (at least for our planet) by proposing that Earth was seeded with biochemicals, perhaps even the first living cells, from space. This idea took shape in the 19th century, and has come under closer scrutiny more recently by astrobiologists, who look for life beyond Earth.

There have been several suggestions for the vessels that carried the life-giving materials to Earth. Might encysted bacteria have been able to reach Earth as microscopic dust blasted through the cosmos? Were biochemicals frozen in the ices of comets that vaporized when they impacted Earth? The Philae lander, which analysed comet ice in 2015, would suggest this is unlikely. So perhaps the strongest candidates are meteorites – could biochemicals and even bacteria have been brought to Earth inside the frozen cores of space rocks?

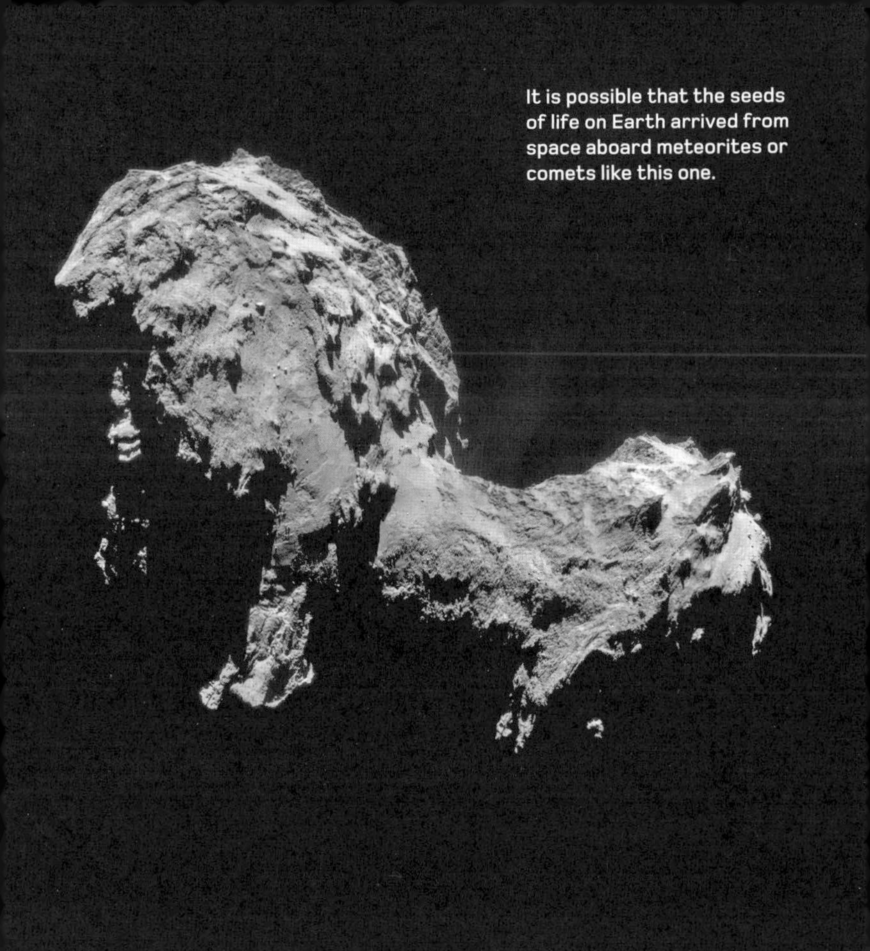

It is possible that the seeds of life on Earth arrived from space aboard meteorites or comets like this one.

Autocatalysis

In recent years, the idea of life cooked up in a warm ancient sea has been pushed aside by an alternative that proposes it arose in hot seafloor sediments. Although extreme to us, the chemical-rich sediments around seafloor hydrothermal vents would have been one of the most stable environments when Earth was young: the surface regions were subject to intense solar radiation and dramatic climate changes. So many believe that the step from non-life to life was made in the ocean depths.

But what exactly transforms a chemical into a life form? The answer is 'autocatalysis', or being a catalyst that makes itself. A catalyst is a substance capable of removing the energy barrier that prevents a chemical reaction occurring – its presence makes the reaction run almost spontaneously. The first life forms were molecules that could catalyse the formation of an exact copy of themselves from a supply of raw materials. RNA is able to autocatalyse in this way, but it is likely that many simpler chemical life forms came before it.

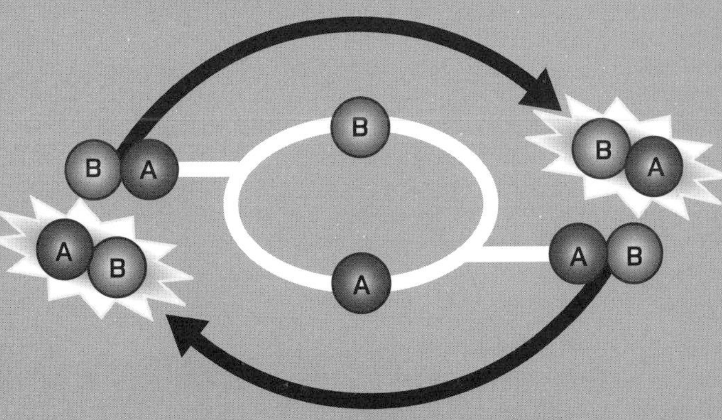

Schematic
representation
of autocatalysis

Chemical evolution

The process of transition from complex molecules to something we would recognize as a life form today is called chemical evolution. We can imagine sediments brimming with chemicals, some of which were able to make copies of themselves through autocatalysis (see page 312). These molecules were in competition for the same kinds of raw materials. Some would have been better at building accurate copies than others, and so they multiplied and became dominant – the first form of natural selection.

As the replicators became larger and more complex, copying errors or mutations would appear, helping or hindering each molecule in its battle. At some point, a replicator similar to today's nucleic acids (RNA and DNA) associated with proteins, which helped in its quest to copy itself and protect the delicate molecule. The proteins were coded into the structure of the replicator, creating a life form akin to a virus (see page 342). The final step saw the whole assemblage shrouded in an oily membrane to protect its supply of materials – the very first cell.

Endosymbiosis

The first cellular life form to evolve out of the chemical stew around 3.5 billion years ago was the ancestor of the prokaryotes, the Bacteria and Archaea domains that still dominate life on Earth. The third domain of life, the Eukaryota, descend from a later single cell that evolved about 1.5 billion years ago. And surprisingly, this cell did not have one ancestor but several. Some of its organelles (see page 60), such as the endoplasmic reticulum, evolved from folds in the cell membrane, but theory suggests that others – the crucial mitochondria and chloroplasts – came from prokaryotes living in symbiosis inside a larger cell. It is possible that this process of 'endosymbiosis' occurred many times, but all of today's eukaryotes are descended from a single victorious cell. The earliest endosymbiont was probably a sulphur-eating bacteria that evolved into mitochondria (these organelles still carry their own supply of DNA left over from when they lived free). Chloroplasts probably joined later, originating as independent photosynthetic 'cyanobacteria'.

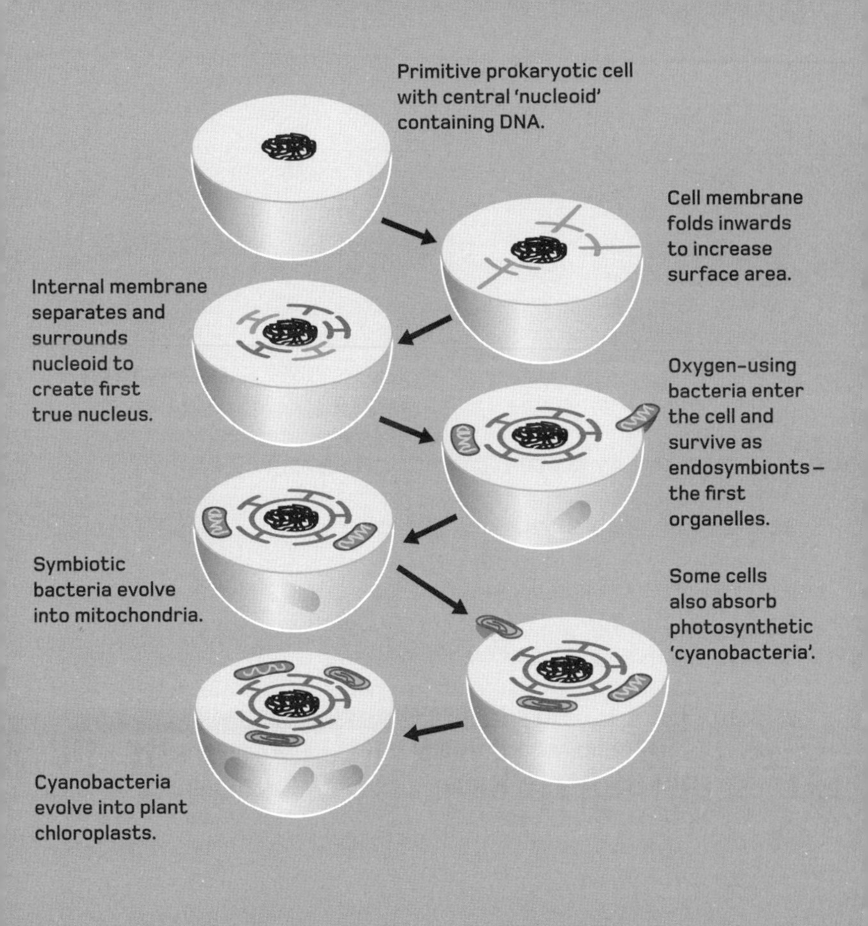
Primitive prokaryotic cell with central 'nucleoid' containing DNA.
Cell membrane folds inwards to increase surface area.
Internal membrane separates and surrounds nucleoid to create first true nucleus.
Oxygen-using bacteria enter the cell and survive as endosymbionts – the first organelles.
Symbiotic bacteria evolve into mitochondria.
Some cells also absorb photosynthetic 'cyanobacteria'.
Cyanobacteria evolve into plant chloroplasts.

Evo–Devo

Short for 'evolutionary development biology', evo-devo seeks to bridge the gap between differences in DNA (genotype) and anatomy (phenotype). In doing so, it is a powerful tool for finding the major branches of the tree of life, helping to fill in vast gaps in the fossil record. Instead of simply comparing DNA or anatomy, evo-devo compares the way organisms develop from a zygote, through an embryo into a mature form (see page 154), and assumes that organisms that develop in the same way early on are more closely related than those that follow a different path.

Evo-devo has been at the forefront of evolutionary biology for the last 30 years, and has shown that body complexity and gross anatomical features are not necessarily good indications of an evolutionary relationship. For example, flatworms and segmented worms are related to molluscs, while crustaceans and insects are more closely related to roundworms (nematodes). Meanwhile, starfish have turned out to belong on the same branch of the tree of life as vertebrates.

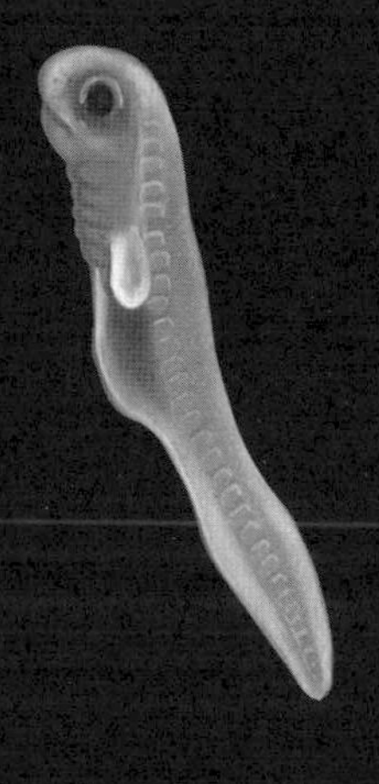

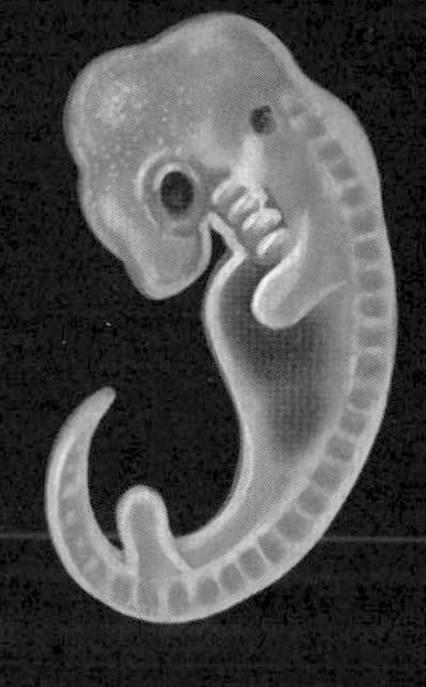

In the early stages at least, all vertebrate embryos develop in the same way.

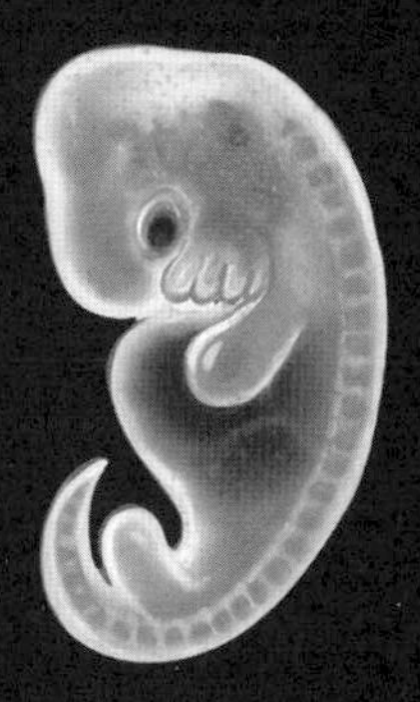

Hox genes

A prime example of the power of genetics, Hox genes are the 'source code' that controls embryonic development in most animals Often characterized as a 'developmental-genetic toolkit', they are used in all Bilateria, the 'subkingdom' of animals that grow bodies with bilateral symmetry at some point in their lives. Thus, Hox genes are shared by organisms ranging from flatworms to fin whales.

These genes produce an embryo with a head at one end and an abdomen (generally with a tail-like structure) at the other. There are genes for every body segment in between, tagging different parts of the head-tail axis with specific proteins. These chemical tags result in the right anatomical feature growing at each location – perhaps a limb, other appendages or an eye. Hox genes are so important that natural selection has left them more or less identical across the Bilateria, performing the same role everywhere. Some may be supressed in certain species – such as limb genes in snakes – but the basic body plan is always retained.

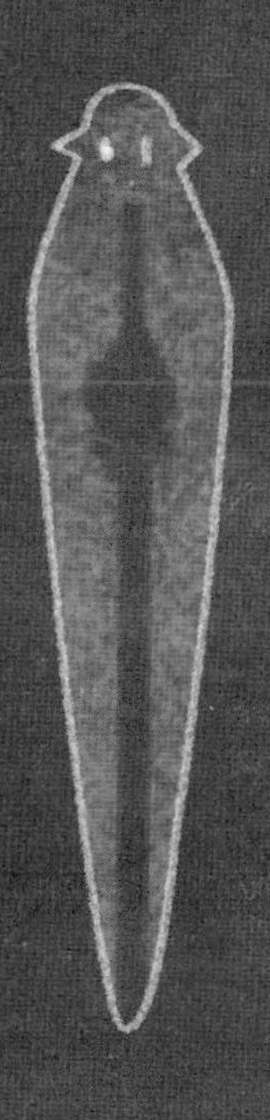

Flatworm

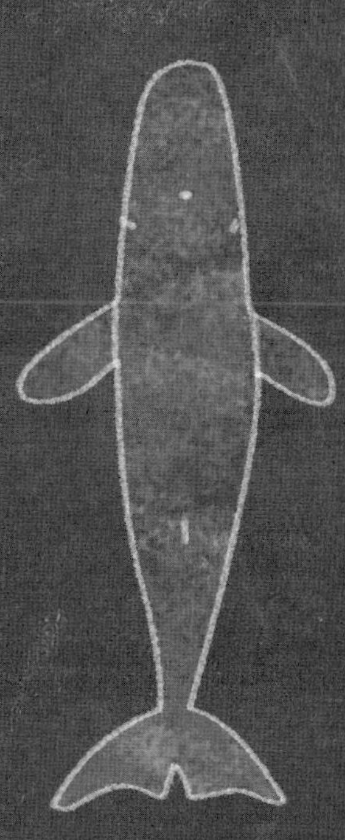

Fin whale

Humans and genetics

One of the main reasons we seek to understand genetics is so that we can use it to solve problems, not least problems with the human body. Many inherited disorders that cause great suffering are caused by genes, and could perhaps also be cured with genetics. The likes of sickle-cell anaemia, haemophilia and Huntington's disease are caused by a single gene, while others are the product of larger errors at the level of the chromosome. Many diseases have genetic components as well, and the future of medicine is likely to be focused on tailoring treatments individually, according to the patient's genetic makeup.

Human genetics is best expressed using the 'karyotype'. This is a snapshot of all of a person's chromosomes as they are coiled up ready for a cell division. There are 46 in all: one pair of sex chromosomes and 22 pairs of autosomes. This is the first place doctors search when diagnosing genetic disorders, looking for mismatched chromosomes that indicate something is awry.

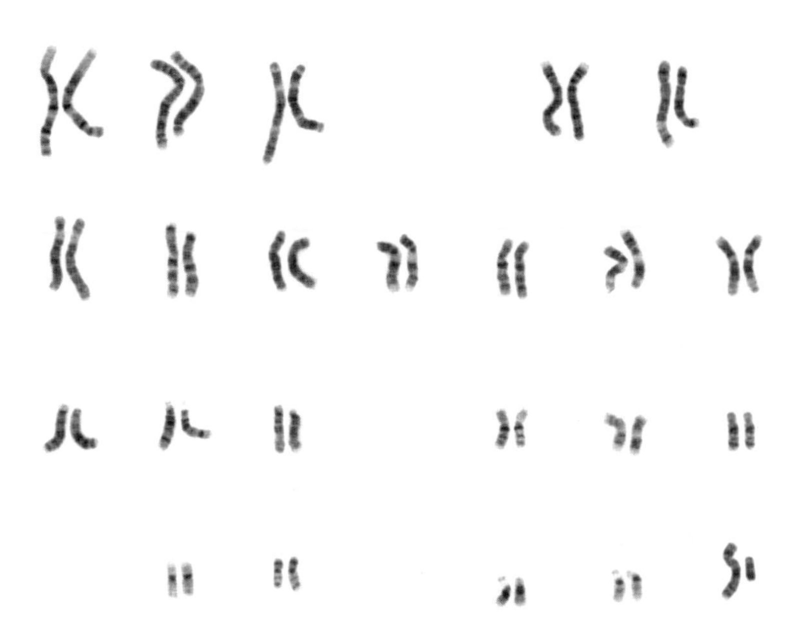

The human karyotype is a representation of 46 chromosomes in 23 pairs.

Human Genome Project

Dubbed 'one of the great feats of exploration in history – an inward voyage of discovery rather than an outward exploration of the planet or the cosmos', the Human Genome Project was 20 years in the making. In 2003, it succeeded in reading every piece of human DNA, sequencing the nucleotide bases in the sense strands of the haploid human genome (that's 23 chromosomes, not the full 46). The Human Genome Project got most of its source material from an unnamed man from Buffalo, New York, with a few samples from other donors. The result is 25 lists of nucleotides: 22 autosomes (non-sex chromosomes), both the X and the Y chromosomes, and the tiny strand of mitochondrial DNA. All together, they amount to roughly 3 billion characters, but what does that really add up to? The data amassed by the Human Genome Project is unique to its donors – it is not a blueprint for all humans. But it still provides a reference for identifying those parts of the genome that are coding DNA and those which are 'junk'. In this way, the real exploration of the human genome has only just begun.

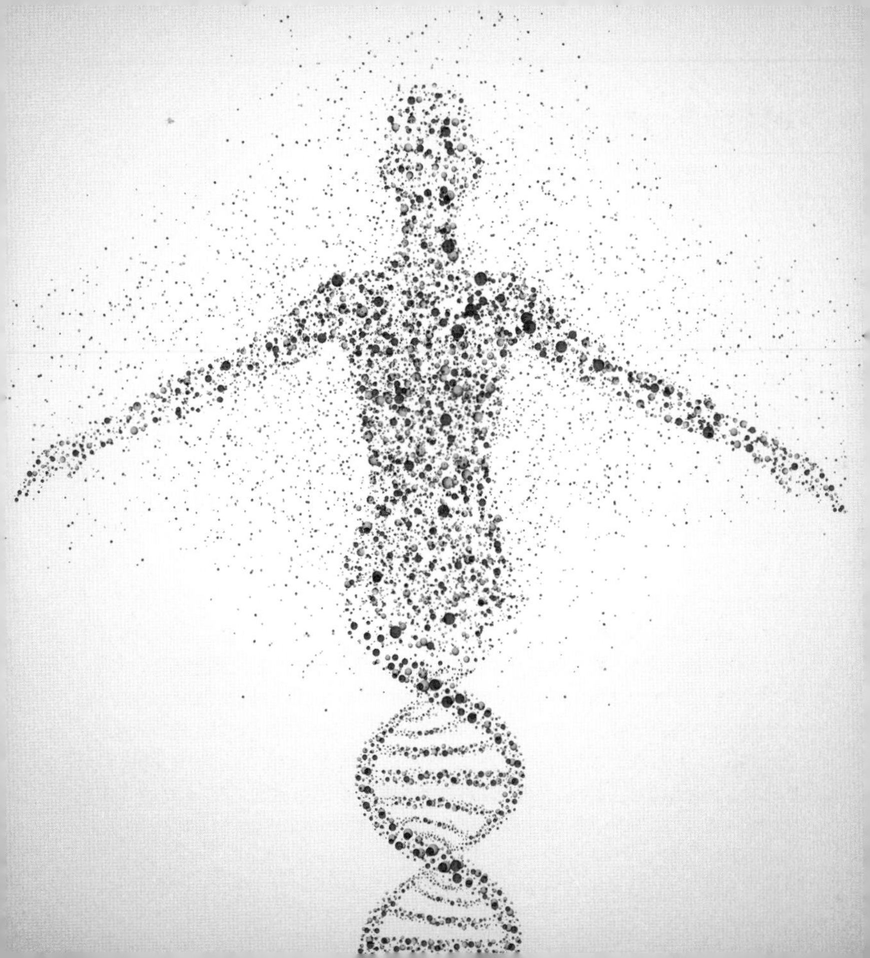

Blood types

Every person has one of four blood types: A, AB, B or O. They are an inherited trait, and the genes involved are an object lesson in Mendelian genetics. The types relate to antigens, or marker chemicals, that appear on the surface of red blood cells. There are also antibodies that roam the blood stream in search of aliens – things with different antigens to the blood. So a person with A blood, has A antigens on their cells and B antibodies in the blood stream. These antibodies lock onto any cells with the B antigen, alerting the immune system.

Blood types are controlled by the genes A, B and O. The genotypes AA or AO result in the A blood type; B blood is from BB or BO genotype. Inheriting both an A and a B allele produces the AB blood group. These blood cells have both A and B antigens, and there are no antibodies. The genotype OO results in the O blood type, where the cells have no antigens and the blood contains both antibodies. This means O blood can enter any system undetected, while AB blood can accept all other blood types.

Blood types

	Group A	Group B	Group AB	Group O
Red blood cell type	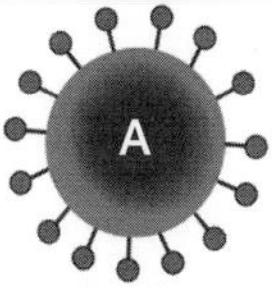 A	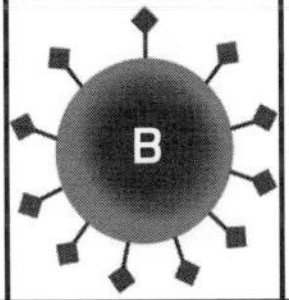 B	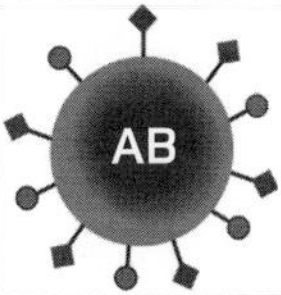 AB	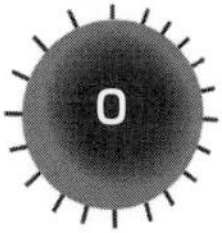 O
Antibodies present	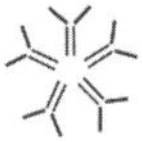 Anti-B	 Anti-A	None	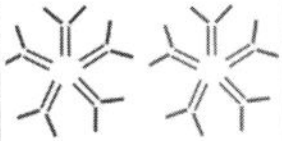 Anti-A and Anti-B
Acceptable donors	A or O	B or O	All	O

HLA tissue types

HLA stands for 'human leukocyte antigens', relating to a set of genes that code for the antigens, or chemical markers, that act as an identity system for the body's cells. The HLA genes produce about a dozen antigens that appear on every cell – some are more important than others. The immune system ignores the cells that have these markers, and attacks anything that has an alien antigen. This is how pathogens, or infectious agents, are generally identified for removal. (Some pathogens and parasites are able to adopt the HLA of their host and so stay hidden.) An individual's HLA types can be ascertained using antibody tests in the lab. This is how donor organs are matched to recipients for transplant. HLA types are also linked to ethnic groups, and are useful in research into human migration. Most of the HLA genes are clustered on chromosome 6. Some of them are associated with inherited disease such as coeliac disease and types of arthritis. This is because the disease genes sit right next to the HLA ones on the chromosome and are likely to be inherited together.

HUMAN
ORGAN
FOR TRANSPLANT
KEEP
REFRIGERATED

Race

In the 19th century, one of the goals of the emerging field of anthropology (the study of humans) was to put our species into its biological context. One of the results was the notion of racial groups, frequently simplified into a few major races such as australoid (from Australia), mongoloid (East Asia and Americas), caucasoid (Europe and South Asia) and negroid (Africa). Whatever its origin, this kind of thinking was frequently deployed as a way of proving the superiority of Europeans. Attempts were made to link inherited phenotypes such as skin colour, hair type and skull shape to intelligence, personality and even morality. All of them failed, but that did not stop pseudoscientific opinions taking hold – the impact of which linger to this day. In population genetics, the term 'race' is seldom used. Instead a gradual change in the phenotype of a population across a wide region is called a cline, and this better reflects the many phenotypic differences of humans. Nevertheless, the differences between all 7.3 billion of us comes down to just 0.5 per cent of our DNA.

Polyploidy

It is not unusual for plants, especially crops, such as wheat and tomatoes, to carry multiple sets of chromosomes over and above the normal two. This phenomenon is called polyploidy, and in plants it results in larger plant bodies – hence its appearance in crops. Some animals are polyploid with no ill effects – a few fish species have as many as 400 chromosomes in every cell. Polyploidy is often linked to parthenogenisis (see page 160) in which a female produces young without needing to mate. This carries a higher chance of cell division errors placing multiple sets of chromosomes into a zygote. A similar process can happen in humans, most commonly when an ovum (egg cell) is diploid and already contains 46 chromosomes. The sperm adds another 23 at fertilization, resulting in a 'triploid' (three-set) zygote with 69 chromosomes. It is estimated that 2 per cent of all human conceptions result in triploidy – plus a few more producing 'tetraploidy'. The great majority will result in miscarriage, with 15 per cent of spontaneous abortions being caused by this single factor alone.

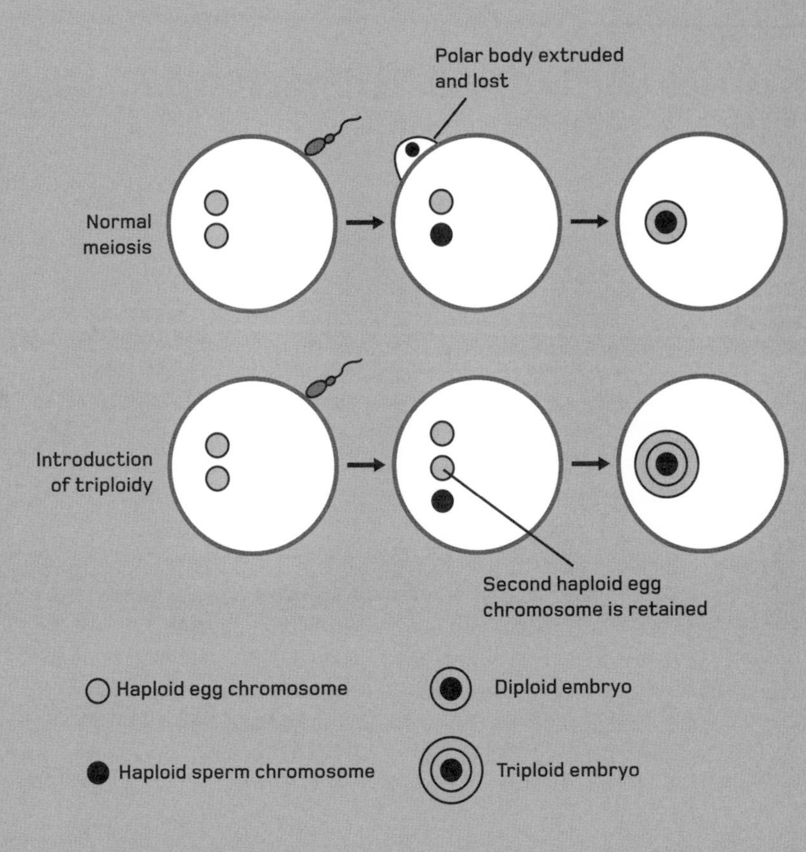
Polar body extruded
and lost
Normal
meiosis
Introduction
of triploidy
Second haploid egg
chromosome is retained
Haploid egg chromosome
Diploid embryo
Haploid sperm chromosome
Triploid embryo

Down's syndrome

Named after John Down, the British doctor who described it in the 1860s, Down's syndrome is the result of a type of chromosomal disorder called aneuploidy. This is when there is an abnormal number of chromosomes in the body cells. In the case of Down's syndrome, the cells are 'trisomy 21', meaning they have three versions of chromosome 21, rather than the normal two. As a result of this extra genetic material, people with trisomy 21 tend to be shorter than average, have distinctive facial features and usually suffer from heart problems. They also tend to have an adult IQ of around 50, in line with the average nine-year-old, although this ranges widely.

Another aneuploidy disorder is Turner syndrome, where the cell has only one X chromosome. Sufferers are female, shorter than average and have fertility problems. In Klinefelter syndrome, meanwhile, a male has the genotype XXY. The extra X makes him grow very tall and develop a mix of male and female characteristics.

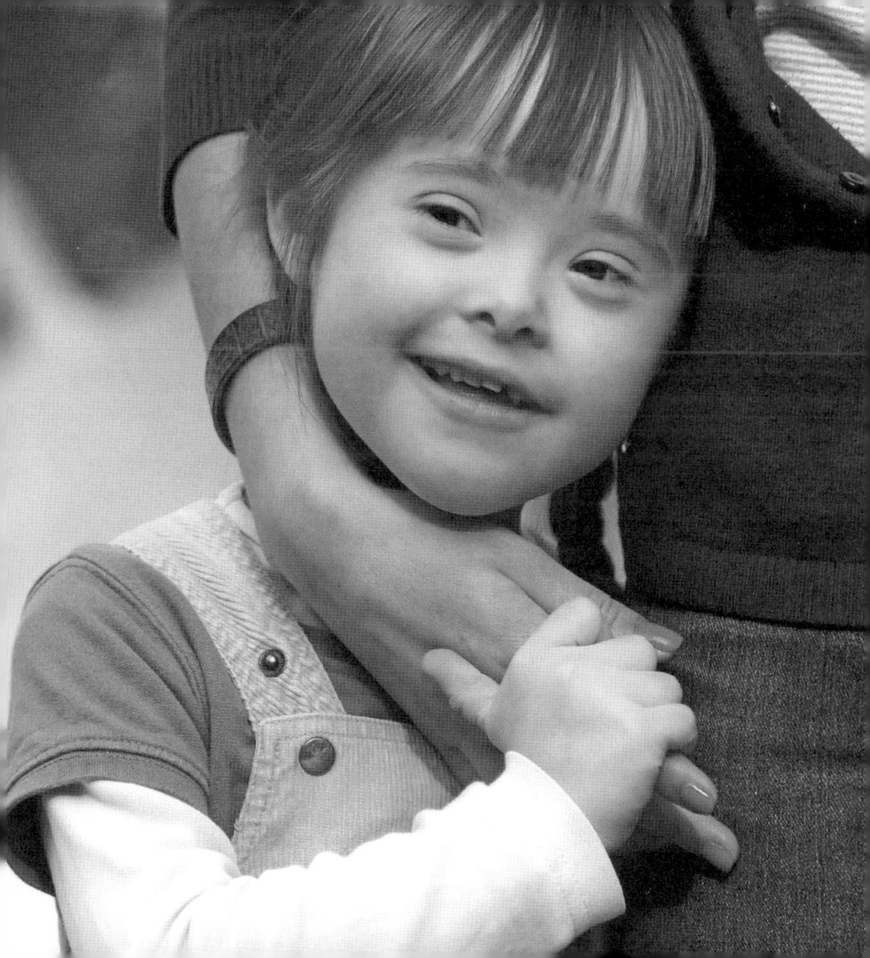

Copy-number variation

A copy-number variation (CNV) is a common chromosomal abnormality. It arises during the replication of a chromosome and results in a sizable section of DNA being deleted or duplicated. As a result, the daughter chromosome has a different number of genes – often multiple copies of the same ones – than in the parent chromosome. Anywhere from one thousand to several million bases can be involved in a CNV. It is estimated that 13 per cent of the variation among human genomes is due to this kind of chromosomal mutation. (Most of the rest comes from point mutations where a single nucleotide base is altered.)

CNVs among the human population were discovered by the Human Genome Project. Stable CNVs, which have little or no impact on the phenotype, are passed on down the generations. Large CNVs can cause infertility, because they produce a mismatch in length of homologous pairs, and this reduces the success rate of meiotic divisions (see page 132).

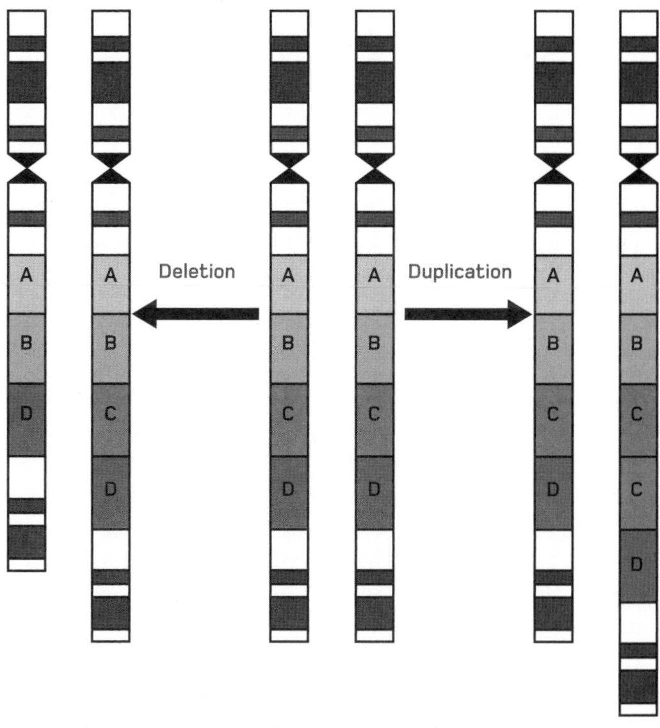

One copy of gene C removed

Original DNA strands with genes A, B, C and D

One extra copy of gene C

X-linked disease

One of the advantages of having diploid cells, equipped with two sets of every gene, is that if one gene proves faulty there is another to override its effects. The 22 non-sex chromosomes or 'automsomes' form matching pairs, but the 23rd pair, made up of the sex chromosomes, can be unequal. Females have two Xs, while all males have an X and a Y. The Y chromosome is much smaller than the X, with 59 million base pairings compared to the X's 153 million. As a result, there are genes present on the X that are not matched on the Y. Therefore, when a deleterious gene appears on an X chromosome in a female cell, the opposing X can mask its effects. Yet that same 'X-linked' gene will be free to express itself in a male cell, since the Y offers no such defence. As a result, several inherited disorders are X-linked and almost exclusive to males. These include colour blindness, haemophilia and Duchenne muscular dystrophy. Females are generally carriers of these diseased genes, but will only suffer themselves if they inherit two X chromosomes carrying the faulty gene.

Tsarevich Alexei Romanov, heir to the Russian throne (second from right), suffered from haemophilia, an X-linked blood disease inherited from his great-grandmother Queen Victoria, who was a carrier.

Cancer

The uncontrolled growth of a body tissue, generally resulting in a tumour, is known as a cancer. The tumour may have far-reaching effects by spreading through the body, disrupting its normal workings and eventually overwhelming a vital organ or pushing the body's immune system beyond its limit.

Cancer is not one disease but many different ones – around 200 in total. They have just as many causes, including exposure to a carcinogenic substance, ionizing radiation, certain infections and also genetics. In most cases it is an accumulation of these factors that trigger development of a cancer, but all cancers begin with a change in certain genes, known as the oncogenes. These genes are involved in rapid cell divisions and are normally switched off after the embryonic stage. However, if they are switched on again, they override the process of controlled cell death that maintains body tissue at a fixed size. The result is uncontrolled growth in a certain part of the body, leading to a tumour.

Every cancer begins with the uncontrolled growth of a primary tumor, arising from a single abnormal cell. Changes in the tumor cells can lead to metastasis, where new, genetically different tumors spread through the body.

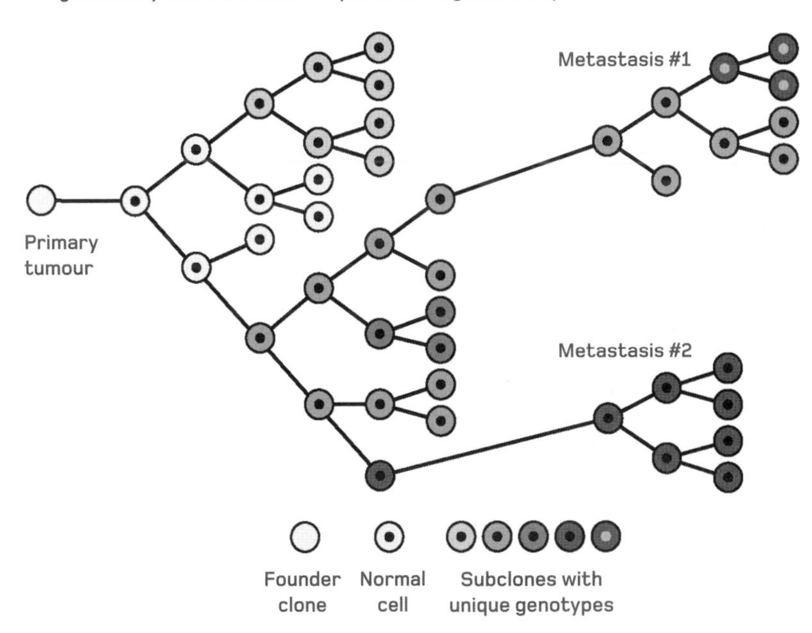

Viruses

Everyone is familiar with viruses – we have all been intimately acquainted with them at some point in our lives when suffering during a viral disease, such as the common cold or chickenpox. But few appreciate that by most measures, a virus is not really a living thing. The best way to understand it is as parasitic DNA.

The virus 'body' is made up of a coil of DNA (or sometimes RNA) surrounded by a protective protein coat. It is parasitic because, unable to replicate its own DNA, it hijacks the replication system of a cell. The protein coat attaches to the membrane of the cell and makes a channel through it for the DNA to enter. The DNA is then taken to the nucleus and causes the cell to make copies of it and its proteins continuously – until the cell is so full that it bursts, releasing new viruses to infect new hosts. This process is what kills cells and creates creates illness. Make no mistake, viruses are no genetic sideshow; a single cup of seawater contains more viruses than there are humans on Earth, each one evolved to parasitise a specific genome—and mercifully few targeted at us.

Viral structures

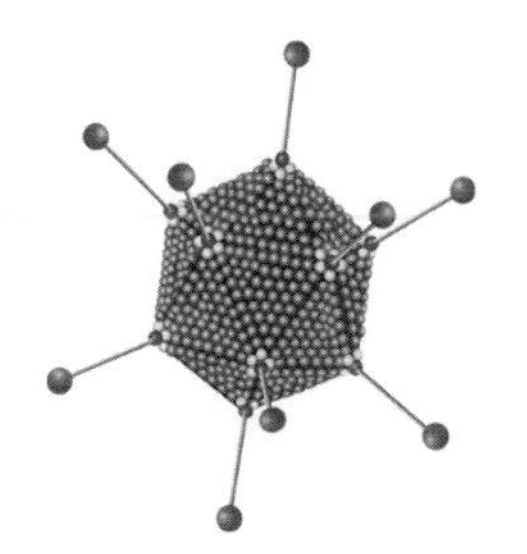

Polyhedral (*Adenovirus*)

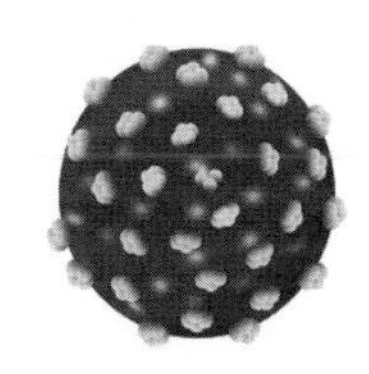

Spherical (*Influenza*)

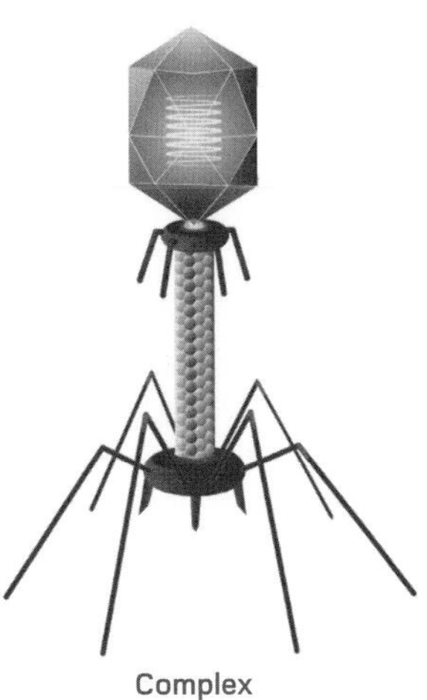

Complex (bacteriophage)

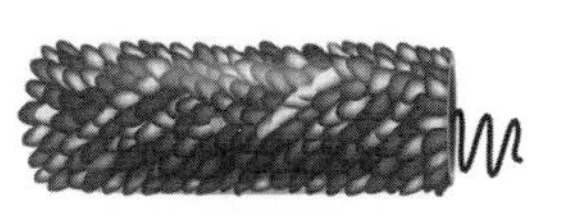

Helical (tobacco mosaic virus)

Prions

Derived from a contraction of the terms 'protein' and 'infection', a prion is a disease-causing agent which, like a virus, is non-living. However, unlike viruses and all other infectious agents, prions contain no genetic material. Instead, they are malformed proteins. They originate as proteins synthesized in the normal way within the cell, and have a structure that makes them of use in metabolism. However, many proteins are able to refold into alternative shapes, of no metabolic value. Prions are a mercifully rare subgroup of proteins that become self-propagating once they become malformed. The misfolded protein acts as a template or mould that causes healthy proteins to take on the same malign shape. Now there are two, and the process continues, creating an exponential buildup of bad proteins. The proteins cluster together to form fibres called amyloids that damage tissue. Prions were not discovered until the 1980s: to date all known prion diseases, such as Creutzfeldt-Jakob Disease, attack the brain or nervous system, have no cure and are invariably fatal.

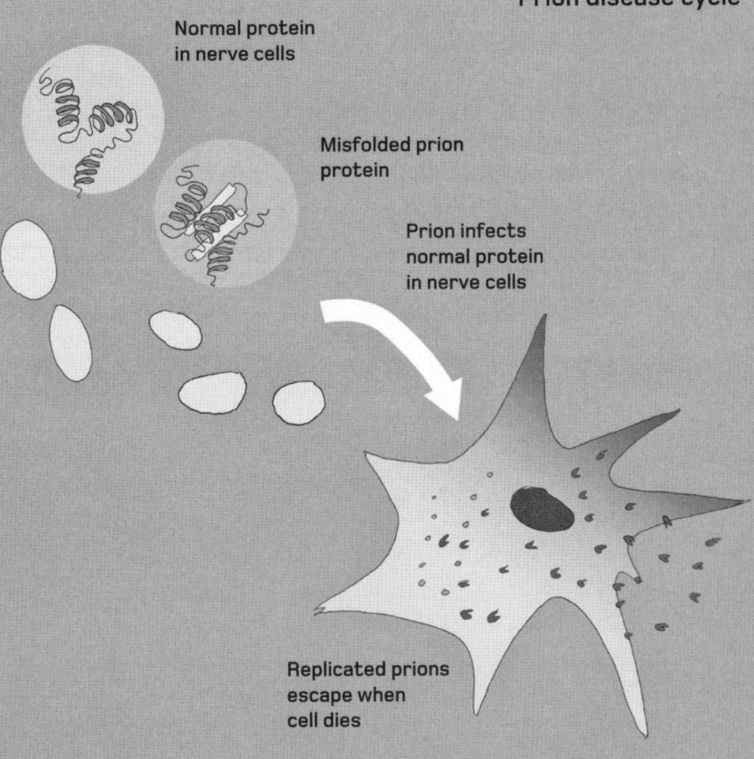
Prion disease cycle
Normal protein
in nerve cells
Misfolded prion
protein
Prion infects
normal protein
in nerve cells
Replicated prions
escape when
cell dies

Nature versus nurture

We have all been drawn into the nature versus nurture debate at some point: does our personality arise from an innate inheritance or is it moulded by our experiences? Most modern thinkers would agree with the teaching of 17th-century philosopher John Locke, that the mind of a newborn is empty of knowledge, a 'tabula rasa', or clean slate. The things we then experience fill our memory and mould our attitudes as we grow, but do genetics play a role? No one suggests that states of mind and social attitudes are inherited. However, the fabric and function of the brain that creates those states probably is.

In the United States, defence lawyers can argue that the structure of their client's brain reveals a cognitive deficiency to explain and excuse a crime. However, studies of brain development during pregnancy and after trauma show that it is 'plastic' and alters its functional map throughout life. This would suggest that in most cases, nurture tends to dominate nature.

Eugenics

Meaning something like 'well born', eugenics is now seen as embodying the darker side of human genetics, a theory that the human species could be improved by breeding out unwanted personality traits in favour of intelligence and other desirable qualities. The idea was the brainchild of Francis Galton, a cousin of Darwin. Being a Victorian gentleman, he assumed that his kind was a superior breed to the rest of humanity, and that the mechanisms of inheritance and selection outlined by Darwin and others could be used to make everyone 'better' – more like him.

But how to identify inheritable traits in the first place? Galton was aware of the nature vs nurture debate (indeed, he coined the phrase), and from the 1880s he attempted to link physical features to intellectual faculties. Galton found no correlations, but the idea of eugenics persisted. Ultimately, it was behind widespread efforts to sterilize the mentally ill (including one Swedish programme that continued into the 1970s), and also found a hideous outlet in the Nazi Holocaust.

Taille 1m
Voûte
Enverg 1m
Buste 0,

Oreille dr. Tête: Long' · Larg' · Long' · Larg'

Pied g.
Médius g.
Auric'r g.
Coudée g.

Coul' de l'iris g.: N° de cl. · Aur'e · Pér'e · Part'és

Agé de
né le
a
dép'
âge app'

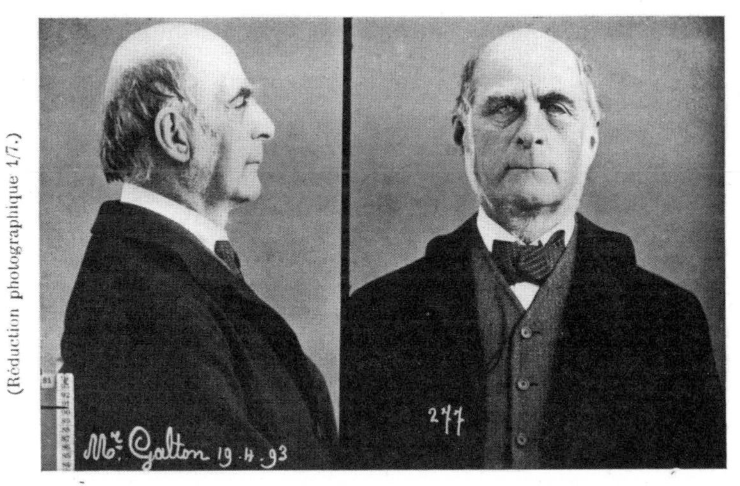

Front: Inclin' · Haut' · Larg' · Part'és

Nez: Racine (cavité) · Dos · Base · Haut' · Saillie · Larg' · l · l · Part'és

Oreille droite: Bord o. s. p. f. · Lob. c. a. m. d. · A. trg. i. p. r. d. · Pli. f. s. h. E · Part.

Barbe
Cheveux
Car

Color' (pig'' · sang'
Ceint.

Autres traits caractéristiques :

Sig' dressé par M.

Francis Galton used himself as a benchmark in the search for a link between body shape and intelligence.

Intelligence and IQ

There has long been an assumption that intelligence is inherited. An early idea was that brain and skull size correlated to intelligence. All attempts to prove this failed, not least because there was no way of measuring intelligence. In the 1890s, French neuroscientist Alfred Binet took a new approach by devising a test for intelligence, based on problem solving while avoiding the need for advanced reading and writing skills. Binet's test questions got progressively harder, with each one designed to be answerable by 50 per cent of a specific age group, rising with each question. The place in the test where a person faltered showed their 'mental age'. From 1916, a similar test that produced an average score of 100 became known as the IQ test, measuring a person's 'intelligence quotient'. IQ tests are still around today, although we seem to be getting cleverer: the tests must be regularly upgraded to keep the average score at 100. IQ scores and academic attainment run in families, suggesting they may have a genetic factor. But, the mechanisms by which hypothetical 'genes for intelligence' act remain a mystery.

Twin research

The most powerful research tool into the genetics of personality and intellect is twin research. Identical twins are physically and genetically the same in every way – but while their nature is a match, what about their nurture? The holy grail of twin research is to study identical twins that have been separated soon after birth. The twin's nature might make them grow up into adults that act the same and share the same likes and dislikes. But if the way they were nurtured is a major factor, then these identical twins need not share a similar personality. Such research offers the prospect of discovering 'a gene for' all kinds of mental traits.

In practice, separated twins frequently present similar personalities, and statistical analysis suggests about half of that similarity is due to genetics. Isolating an actual gene that codes for personality has proved impossible to date, but a better understanding of the way a brain develops under various stimuli may eventually point the way to the genetic component.

Epigenetics

The term *epi-* means 'on top of', and an epigenetic process is one that impacts on the expression of a gene due to some environmental factor. Multigenerational studies have recently made a rather startling discovery: epigenetic effects seem to be passed on along with genes, at least for one or two generations.

As well as a genome, every cell carries an epigenome – an array of helper chemicals studding the chromosomes. Some helpers coil up unused genes to save space (for example, blood cell genes inside a bone cell). Others unravel sections that are in constant use. Unlike the genome, the epigenome changes in response to environmental stimuli, and researchers are racing to find out more, acting on the hunch that this process is what links diseases such as cancer with poor diet and other bad health choices. Further to this, evidence is growing that the epigenome– or at least some of it – can pass to the offspring, and even to grandchildren. That would mean our genetic inheritance is somewhat pliable, and not set in stone.

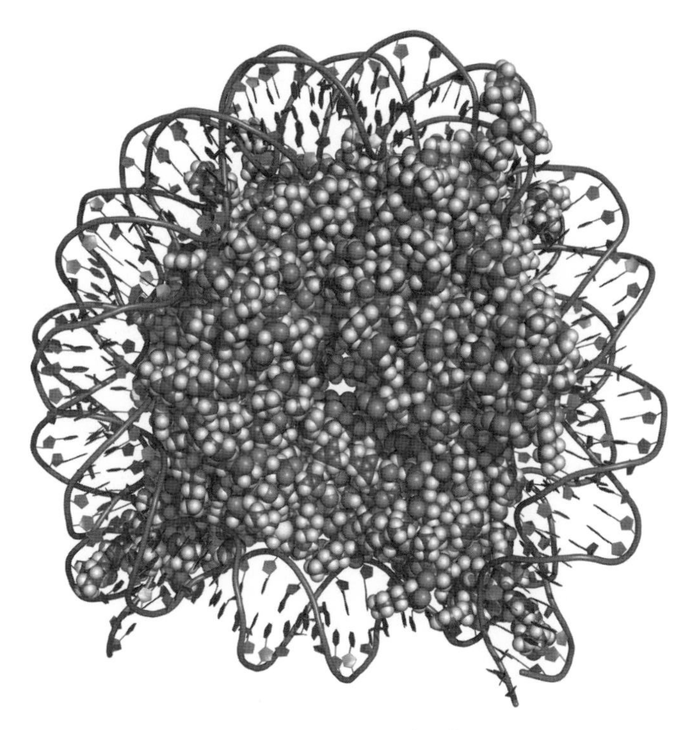

DNA is bound together by a protein core in the chromosome. Proteins and other molecules form the epigenome, which presents certain genes for use and hides others away.

Dutch Hunger Winter

The first evidence of epigenetic inheritance emerged from the Dutch Hunger Winter, a famine caused by a Nazi blockade in the winter of 1944–45. It resulted in thousands of deaths from starvation, but after the war it presented a unique opportunity to study the effects of malnutrition. Babies conceived before the famine had low birth weights; they lacked nutrition in the final stages of development and so did not grow much, remaining small throughout their lives. Babies conceived during the famine, however, had normal birth weights: their early development was during the famine, but the final trimester occurred after it, and their growth caught up. In later life, however, this second group were found likely to be obese and suffer mental illness – and surprisingly, so were their children. The theory is that such problems are caused by an epigenome, created by the famine, that formed in the mother, the foetus and the foetus's germ line (which would eventually produce its own gametes). The question remains – can the epigenome pass further, beyond these generations?

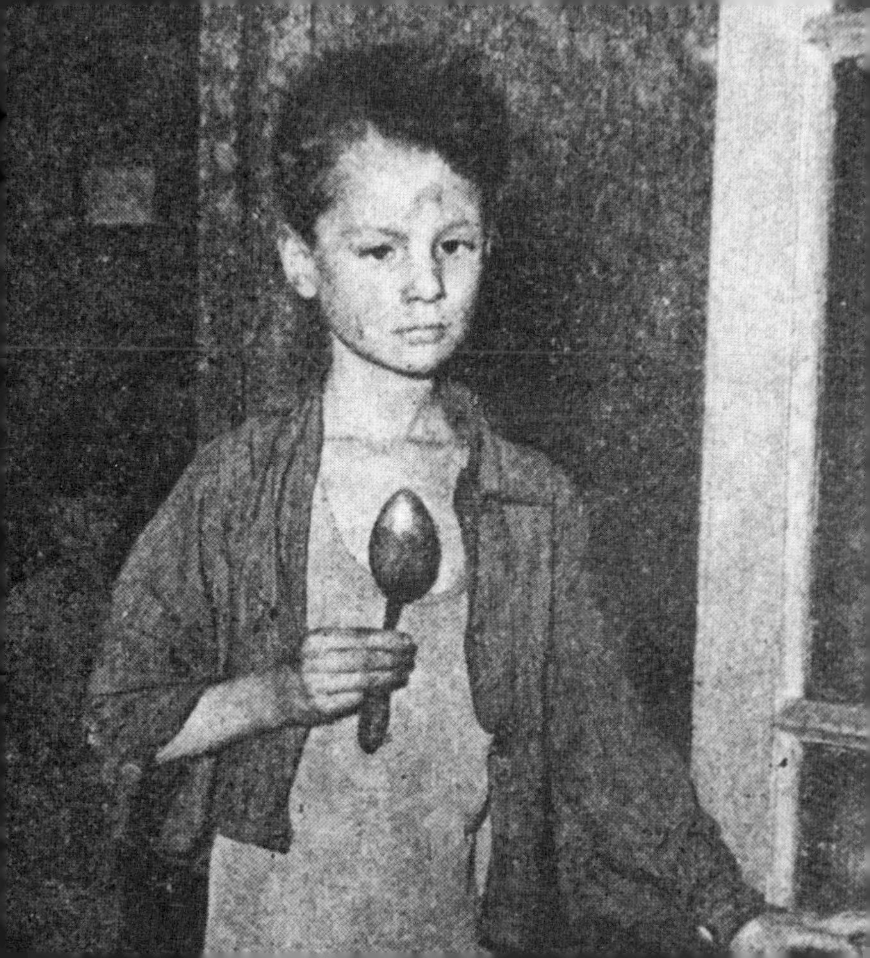

Memetics

Are genes the only things that are subject to natural selection? One proposal suggests not. Memes, a unit of knowledge or memory, can pass from mind to mind replicating in the same way as a gene would do. For example, the meme for applause is a highly successful one. We learn, or inherit, this idea from others, and pass it on in the same way. Clapping has remained remarkably stable for years and is used across most cultures. However, there are 'mutant' forms, where applauders stamp or bang a table. These mutants have taken root in habitats where they fit better than clapping. Some memes are less successful, only spreading among certain communities or being completely forgotten – effectively going extinct. Memetics seeks to use genetic motifs to investigate the nature of ideas, but it eventually fails. While our definition of the meme equates to that of the phenotypic gene, it does not match the genotypic one, a physical strand of DNA. Ideas are not stored by the brain as discrete memes, but are recalled by associating a complex and distributed set of different memories.

Genetics and technology

Gregor Mendel (see page 32) discovered genetics because he chose to investigate heredity in domestic plants and animals. (Before settling on peas as his subject, he bred some terrifyingly aggressive bees that had to be exterminated.) From the very beginning, genetics was related to practical applications, and with recent advances in genome research and genetic engineering, that link is stronger than ever.

Research into stem cells and gene therapy offers the very real prospect that inherited disorders can be fixed at the genetic level and once-permanent injuries could be healed. Genetic modification allows genes from one species to be transferred to another, bypassing the normal rules of breeding. While such a technology (like any kind) has to answer many ethical concerns, it has the potential to transform agriculture – and even humanity itself. Additionally, the chemical properties of DNA are being explored – not for their role in inheritance – but as a material for making machines on the nanoscopic scale.

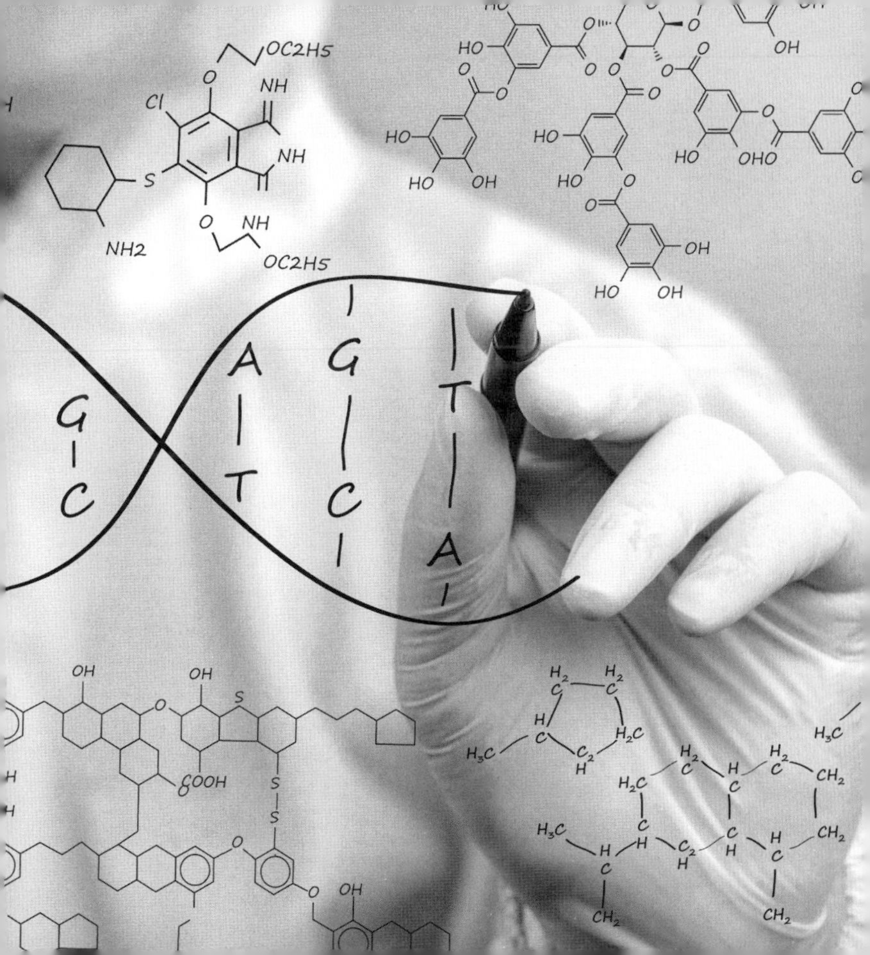

OC2H5
Cl
NH
NH
S
NH2
NH
OC2H5
HO
HO
OH
HO
HO
OH
HO
OHO
OH
HO
OH
A
G
T
G
C
T
C
A
OH
OH
S
COOH
S
S
OH
H_2
H_2
H_3C
H_2C
H_3C
CH_2
CH_2
CH_2
CH_2

Artificial selection

All crop plants and farm animals, and most pet animals, are the product of 'artificial selection'. This process makes use of the same inheritance mechanisms as natural selection, but instead of the environment selecting the individuals that live or die, and which pairs get to mate, this is the decision of a human breeder. The breeder chooses individuals from one generation that have certain desirable traits, and mates them with each other in the hope that the traits will become blended together in their offspring. It can be a rather hit and miss affair – as Mendel found, not all offspring will express the targeted traits. Those that do not are excluded from future matings, but even so, artificial selection takes many generations to have a noticeable impact.

Artificial selection was the first form of genetic technology although it was practised long before the rules of genetics were revealed. Nevertheless, over the centuries it has produced many of the our most familiar plants and animals.

68
60

Agriculture

Around 12,000 years ago, human society began to transform from one based on hunting and gathering food, to one that grew its own supply. The birth of agriculture went hand in hand with a number of technologies – the plough, irrigation and artificial selection – that took control of crop phenotypes to maximize harvests. This last step is of particular interest. The first farmers are thought to have grown fields of grass that later became today's cereal crops. Their ancestors were no stranger to these foods, collecting the seed grains that fell from wild grasses for grinding into meal. The ears, or fruits, of wild grasses shatter, dropping ripened grains to the ground at the slightest touch, and maximizing their chances of finding fertile soil. However, some grasses do not shatter so easily: natural selection should have selected against these plants, but early farmers realized they were easier to harvest, and grew them all together in the first fields. Today, these same mutant grass strains, which would not thrive in the wild, are among the world's most common plants.

Livestock

Many of the most common and familiar animals tend to be livestock – domestic animals that are raised for their meat, or perhaps eggs, milk, hair or skin. Similar working animals have been bred over many generations to suit human requirements. However, they all have a wild origin and many of their features can be traced to their free-living forms.

Sheep and goats are bred from desert-living mountain animals: they are able to survive in arid climates unsuited to other grazers, and herd together for protection – domestic flocks, like their wild relatives, still run up slopes when threatened. Chickens, meanwhile are domestic cousins of the Indian jungle fowl, a ground-living forest bird that is only capable of short flights and is therefore easily managed. Horses, meanwhile, are descended from fast-running grazers whose hierarchical social groups enable their domestic descendants to work well with a human trainer. Finally dogs, probably the earliest domestic animals, are tame wolves – pack animals that have merged with human families.

Hybrid vigour

Often, the animals produced by deliberately crossing two very different kinds of parent not only have the traits selected by the breeder, but are also strong and healthy – a phenomenon known as hybrid vigour or heterosis. This is an example of the beneficial effects of outbreeding – the mating of organisms with widely different genotypes. The result of such unions is offspring that have the benefit of many different alleles that generally made their parents fit individuals, and winners in the continued race against predators and parasites. One of the benefits of sexual reproduction is that it promotes outbreeding, although there can also be drawbacks – occasionally, offspring inherit incompatible alleles, reducing their fitness.

The opposite of outbreeding is inbreeding, where closely related individuals mate. They share a lot of the same genes and as a result deleterious recessive alleles that would be masked by outbreeding appear in the phenotype more frequently, creating a less fit individual.

A thoroughbred racehorse is the product of hybridizing sturdy English hunting horses with fast and spirited Arabian breeds.

Artificial hybrids

Animal breeders have found they can cross closely related species to produce artificial hybrids, the most familiar of which is the mule, a much-valued beast of burden that is big and strong, but also docile and rugged. A mule is a hybrid of donkey and horse, specifically a male donkey and a female horse (a female donkey and male horse produce a hinny, which is generally smaller and weaker.) A horse has 64 chromosomes while a donkey has 62. As a result a mule has 63, an odd number that makes it very unlikely that the mule can pair up its chromosomes during meiosis and produce viable sperm and eggs.

Other artificial interspecific hybrids include the zonkey (donkey–zebra), beefalo (cow–bison), wholphin (killer whale–dolphin) and pumapard (puma–leopard) The liger (opposite) is a cross between a female tiger and male lion. It has a blend of the tiger's stripes with the lion's paler coat –and thanks to hybrid vigour, it is huge, growing to 3.6 m (11.8 ft) long, which makes it larger than any wild cat species.

Mutant research

Few livestock breed more than once a year, and the long generation times, combined with obvious ethical issues, mean that it is not often possible to research the impact of mutant genes on their fitness and embryonic development. However, the same problems do not apply to *Drosophila melanogaster*, the common fruit fly. This little insect lives for about a month, and is sexually mature at the age of just 8 hours. It has only three chromosomes (plus two sex chromosomes), and there is the added bonus that fruit flies create giant copies of all chromosomes in their salivary glands, which are easy to analyse.

All of this makes *Drosophila* an ideal species for breeding in large numbers for genetic research. Dozens of mutant strains have been bred, including flies with curly wings, a range of colours for the body and eyes, shorter setae (insect 'hairs'), and even a mutant, known as 'tinman', that does not grow a heart. *Drosophila* is also used in the early stages of researching the link between genes and ageing and brain development.

Genetic modification

Also known as genetic engineering, genetic modification (GM) is the practice of introducing novel genes into a genome –most likely transferring the gene from one species to another. While natural selection can evolve any organism into another – an oak tree into a goldfish or a whale into a fungus – it would take millions (if not billions) of years. GM technology bypasses the rules of inheritance using a number of techniques.

The most simple is the gene gun, an air-powered pistol that fires tiny particles of gold coated with genetic material. The particles are targeted at living cells, obliterating most, but a few will be safely subsumed into surviving cells, with their DNA incorporated into the genome. Another technique is to use *Agrobacterium*, a bacteria that infects host plants with a ring, or plasmid, of DNA (producing tumour-like growths called galls). Genetic engineers hijack the plasmid and use it to introduce new genes (that don't make galls), as illustrated opposite. Finally, GM can also use reengineered viruses to inject DNA into cell nuclei.

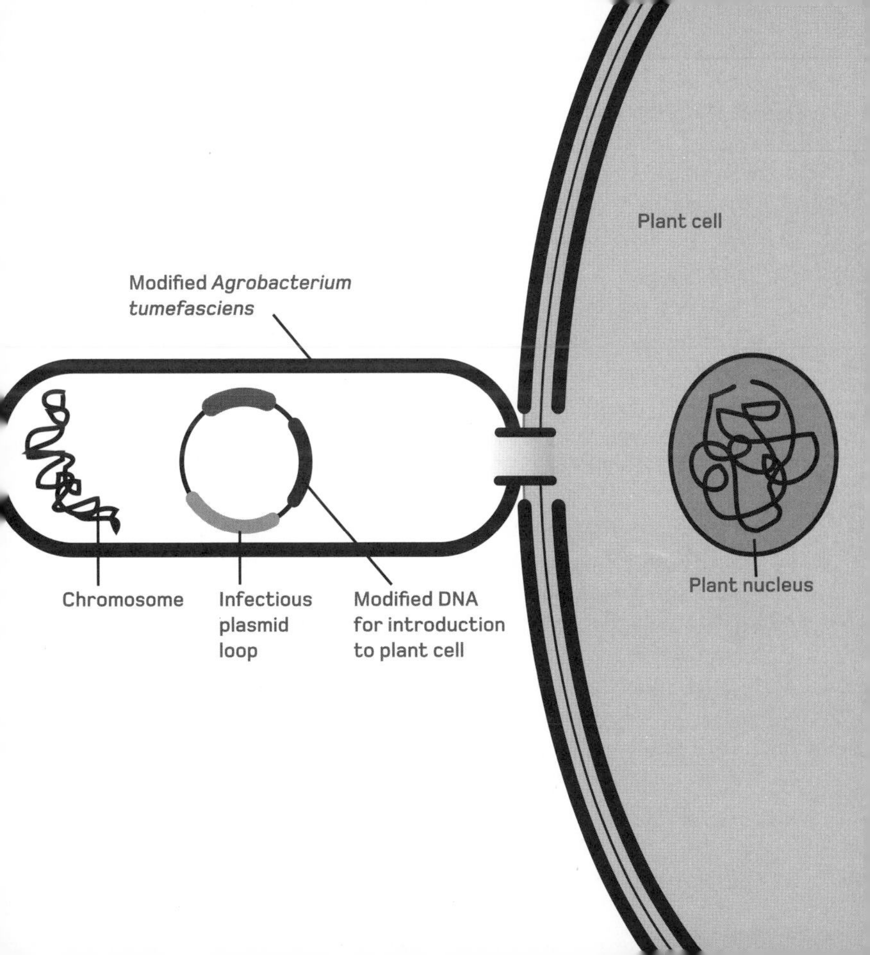
Plant cell
Modified *Agrobacterium tumefasciens*
Chromosome
Infectious plasmid loop
Modified DNA for introduction to plant cell
Plant nucleus

GM food

There have been dozens of attempts to create genetically modified foods – almost exclusively plants. Many have been a failure, either because the genetic modifications have not offered any benefit or because they offer a novelty that has not been met by public demand. One notable failure is the 'fish tomato' given a gene for the antifreeze protein used by an Atlantic flounder in the hope of producing a frost-resistant plant.

Several GM varieties of crops – including pineapple, courgette and potato – have been given genetic resistance to viruses, but the most widespread GM foods are maize (corn) and soya, modified to tolerate pesticide chemicals. But even the use of successful GM foods are heavily controlled in most countries. And more questionably, GM technology has also created a 'terminator gene' that prevents GM crops from setting seed. This would force farmers to always buy new supplies of seed for each season, although there is currently a worldwide moratorium on using the terminator gene in crops.

GMOs

Short for 'genetically modified organism', GMOs include more than just GM crops. Many animals have been genetically modified for reasons other than agriculture – often outlandish ones. Some of the most successful are GM strains of *Escherichia coli*. Often associated with deadly food poisoning, *E. coli* has also been engineered to produce a variety of medicinal substances including the insulin hormone used by diabetes sufferers, growth hormone for treating dwarfism, and clottng factors crucial to the wellbeing of haemophiliacs.

Other GMOs are used as test beds for new medical treatments. They include mice that are modified with the bioluminescence gene of a jellyfish. These rodents literally glow in the dark. Another strange GMO is the 'spider goat', which has the gene for spider's silk incorporated in its genetic recipe for milk. Liquid silk proteins are produced in large quantities in the milk – far more than could be harvested from actual spiders – and can be used to investigate this incredible substance.

Gene patenting

Genetic modification is big business that requires huge outlays in research and development. As a result, the GMOs that result and the techniques used to create them are subject to fiercely defended patents. This means people and corporations actually own specific genes and have rights over every living thing that contains them, a state of affairs that sits uneasily with many people. A patented gene is recorded as a precise sequence of bases, and in order to use that gene – like any other intellectual property – a licensing fee must be paid. This raises both practical and ethical questions. Practically, how does the patent holder tell if their genes are being used? And how do they create a chain of evidence that such a gene got there through misappropriation rather than accidental cross-breeding? Legal wrangles over such issues are now common. More problematically from an ethical standpoint, patents have been applied for on genes for *naturally* occurring substances – including human hormones. In 2013, however, such patents were finally ruled out of order.

Cloning

Clones are organisms that share the same genes: an animal that reproduces asexually is producing clones of itself. Identical twins, triplets and so on, are also clones. Cloning, however, is the technology used to *artificially* create clones from organisms (mostly animals) that normally reproduce sexually to create genetically unique young. An artificial clone is broadly genetically identical to its parent, but they are by no means exact copies – despite what science-fiction authors imagine. For one thing, they are separated by time, with the clone always younger than the parent. They have also developed in a different environment, which may have altered the way they grow. Many animals have now been cloned, ranging from frogs to camels, but many attempts still result in malformations. So why bother with cloning at all? The truth is that it is a powerful tool in a genetic engineer's toolkit because it is the best way of being sure that specific genetic material is passed on unchanged. It is also closely linked to stem cell research (see page 400), where powerful cells can be made to fix incurable ailments.

The Pyrenean subspecies of Iberian ibex became extinct in 2000, but its skin cells have been preserved in the hope of cloning it back into existence.

Nuclear transfer

This form of cloning aims to bypass fertilization and create a zygote directly from an ovum. The ovum has its nucleus removed, along with its haploid set of chromosomes. This process of 'enucleation' is done by hand, using an ultrafine micropipette that can push through the membrane without causing irreparable damage, and leaves the ovum with all of its organelles and other contents intact. Next, the nucleus of a somatic cell with a full complement of chromosomes is put into the ovum. This converts the ovum into a diploid cell, but it is not quite that simple. The ovum's cytoplasm contains elements that are able to reset its new chromosomes, which have been largely turned off in its original cell home. The reset is aided by a pulse of electricity sent through the cell, and there are probably other intricate – and closely guarded – processes used by clone researchers. Once reset, the cell is able to divide and develop towards an embryo that is a clone of the original somatic cell. Clones made this way are mostly used to harvest stem cells, but can be grown into a fully formed animal.

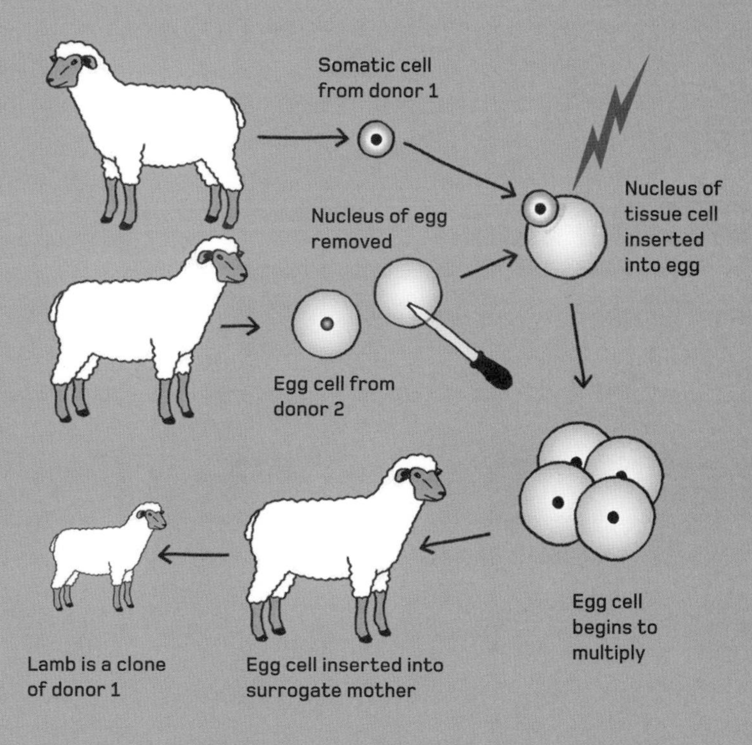

Cloning by nuclear transfer
Somatic cell
from donor 1
Nucleus of
tissue cell
inserted
into egg
Nucleus of egg
removed
Egg cell from
donor 2
Egg cell
begins to
multiply
Lamb is a clone
of donor 1
Egg cell inserted into
surrogate mother

Dolly the Sheep

Perhaps the world's most famous sheep, Dolly was the first mammal to be successfully cloned. Her birth in Scotland in 1996 caused a sensation. Dolly was produced by nuclear transfer: the somatic cell was taken from the udder, or mammary gland, of her mother, and Dolly is named after a country singer famed for the same part of her anatomy.

The nucleus of the cell was placed into an ovum harvested from another sheep, so while Dolly's chromosomes came from the somatic cell donor, her mitochondrial DNA was inherited from her egg donor. Once the transfer was successful, Dolly's zygote was grown to blastula stage (see page 138) in the lab before being implanted into the uterus of a third sheep that carried her to term. (So it could be argued that a clone like Dolly has three parents.) Dolly became an international superstar, but her fame meant she was kept mostly indoors. Most sheep live for around 12 years, but at the age of six, Dolly died from a lung infection, common among sheep living indoors.

Genetic fingerprinting

A fingerprint is a good means of identifying someone: they are effectively unique and when properly analysed, the chances that they point to the wrong person are negligible. The same is true of a 'genetic fingerprint', more correctly called a DNA profile. The profile does not map an entire genome – instead it is a means of comparing two samples of DNA. If a sample from a crime scene matches the DNA of a suspect, it shows that he or she was there.

A similar technique can be used to reveal a genetic relationship between individuals. The system was devised in 1984 by English geneticist Alec Jeffreys to solve a problem: human DNA is 99.5 per cent the same. So to highlight the differences he looked for tandem repeats, places where the same base 'letter' repeated several times. A DNA sample is cut up, and then the chunks with specific repeated sections are amplified, or copied in large numbers. These chunks have a certain length – a feature shared with relatives – and so when the sample is separated by size, it creates a unique pattern that can be compared with others.

Testing suspects

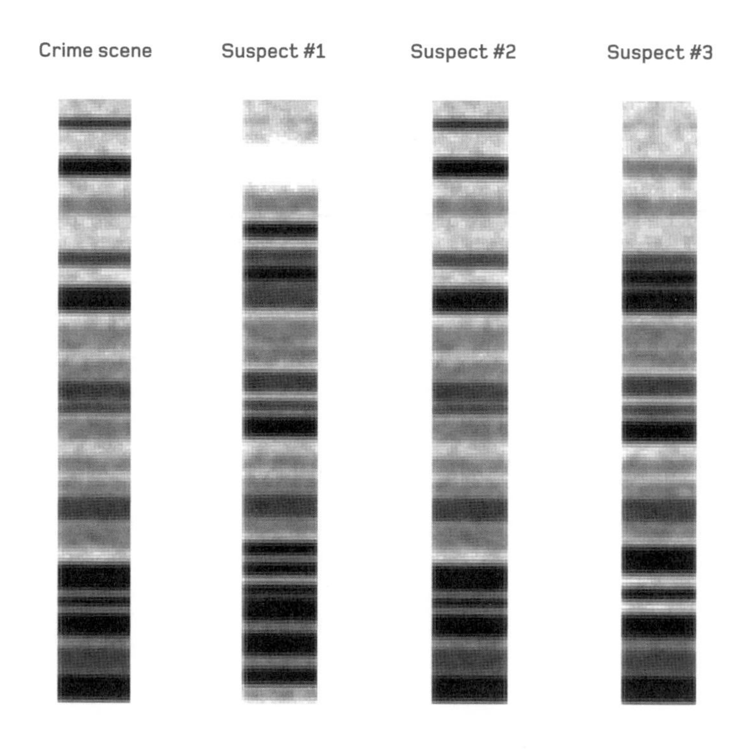

Polymerase chain reaction

Polymerase is the enzyme used in the cell nucleus to read and replicate a single strand of DNA. Geneticists can make use of this copying machinery to mass-produce specific pieces of DNA. The most widespread technique is the polymerase chain reaction (PCR), invented in 1983 and used in genetic profiling, but also to manufacture any large sample of DNA.

The process starts by mixing a piece of target DNA (still within a larger strand) – with polymerase enzymes, a supply of nucleotide bases, and molecules called primers. The primers are short strands of DNA that are coded to attach to the target DNA and mark the point for the polymerase to begin copying. PCR involves a number of cycles, each with three steps. First, the DNA is heated so the helix separates, then the primers are attached and, finally, the polymerases make a copy of the target DNA, plus whatever else is there. The cycle is repeated, making more and more copies – in just 30 cycles (about 4 hours) a single DNA sample can be multiplied into a billion!

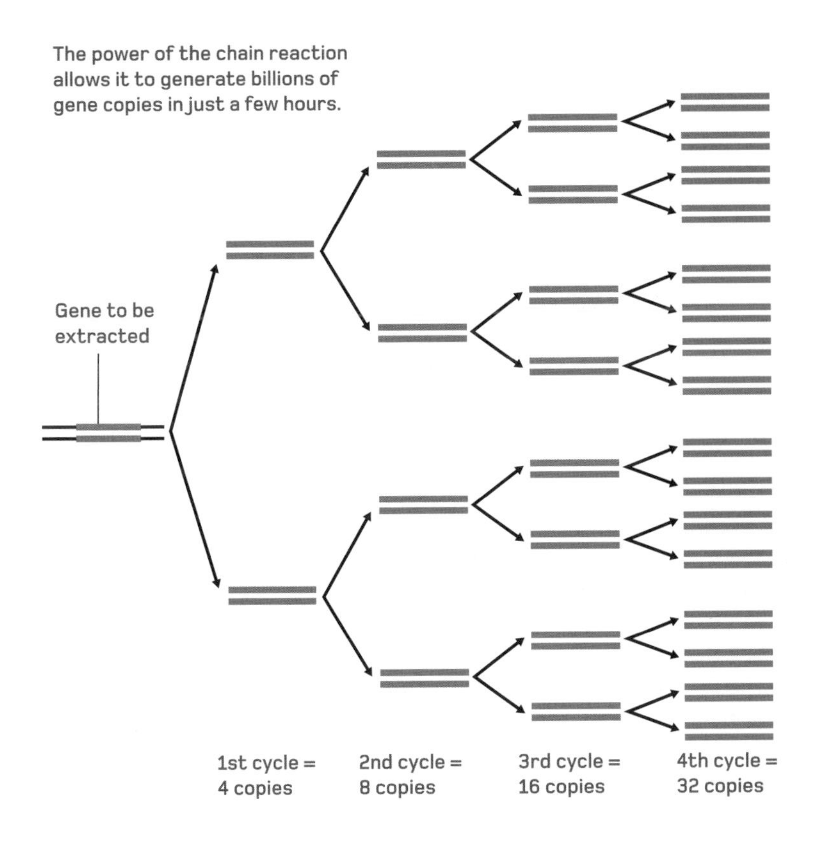

The power of the chain reaction allows it to generate billions of gene copies in just a few hours.
Gene to be extracted
1st cycle = 4 copies
2nd cycle = 8 copies
3rd cycle = 16 copies
4th cycle = 32 copies

Electrophoresis

In order to separate pieces of DNA and other large biochemicals such as proteins, a process called electrophoresis is used. It works by placing the mixture of jumbled DNA and dyes at one end of a plate of gel. (The gel is often made from agar, the jellylike polymer taken from seaweeds). The gel is swamped in a conductive material called a buffer, and an electric current is passed through it. DNA has a negative charge, due to all the phosphate ions involved in connecting up its 'backbone' of sugars, and this means it will migrate through the buffer away from the negative electrode, towards the positive. All the DNA sets off together, making thick bands of dye on the gel, but given more time, these will split into narrower bands, each representing a group of DNA segments with the same number of base pairs. Shorter segments of DNA will travel further than the longer ones. One gel can carry several samples, and eventually each sample creates a particular pattern of bands along the gel (as shown on page 389). These can be used in turn as a DNA 'ladder' – strands of known sizes can be used to estimate the size of bands in other samples.

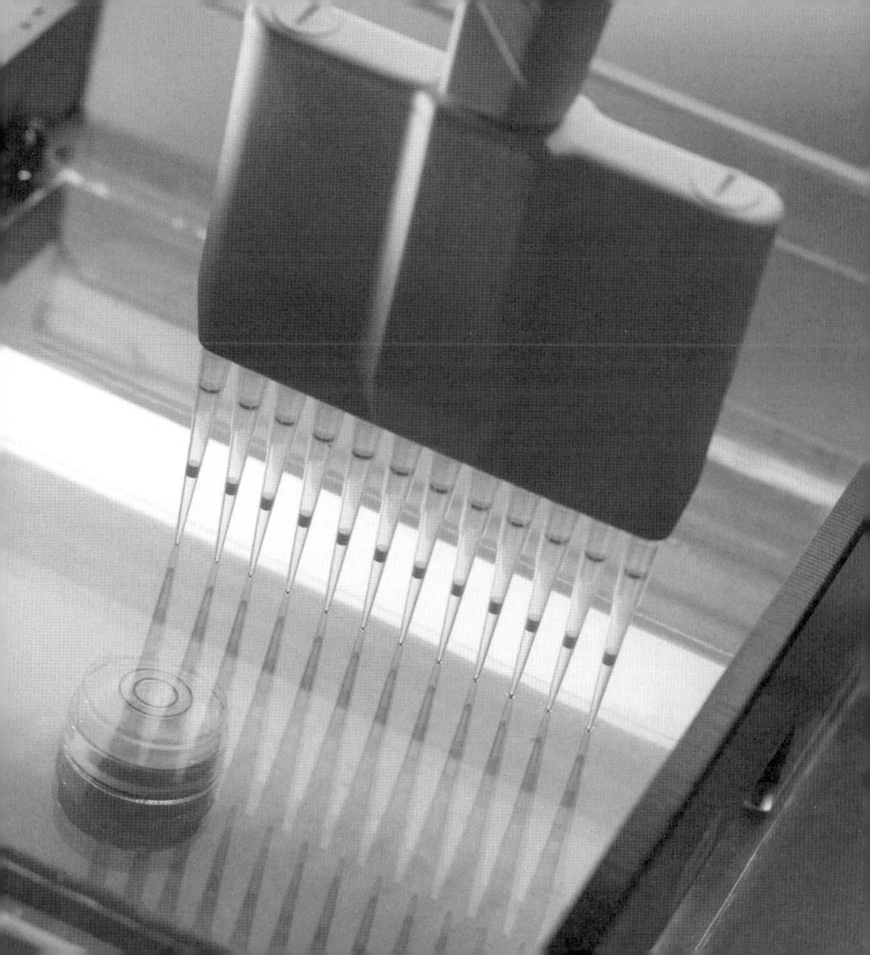

Gene testing

A number of diseases and disorders are strongly linked to inheritance. They include problems like lactose intolerance, porphyria and some forms of Crohn's disease, but the real list is much longer. Medical testing of potential parents has been developed to reveal the presence of these harmful genetic features. In the case of chromosomal disorders such as Down's syndrome (see page 334) the test is a karyotype, which looks at the chromosomes as a whole (see page 322). When a disease is linked to a single gene, then tests will often look for the consequences of that gene. That might be a particular protein or the presence of a metabolite that indicates the gene is at work. Advances in genetics have made it easier, and crucially cheaper, to develop tests for particular DNA codes. But while some genetic disorders can be mitigated with drugs, they are all so far incurable. While a negative result leads to obvious relief, a positive genetic test offers little but anguish. So testing goes hand in hand with genetic counselling to give patients an understanding of the consequences.

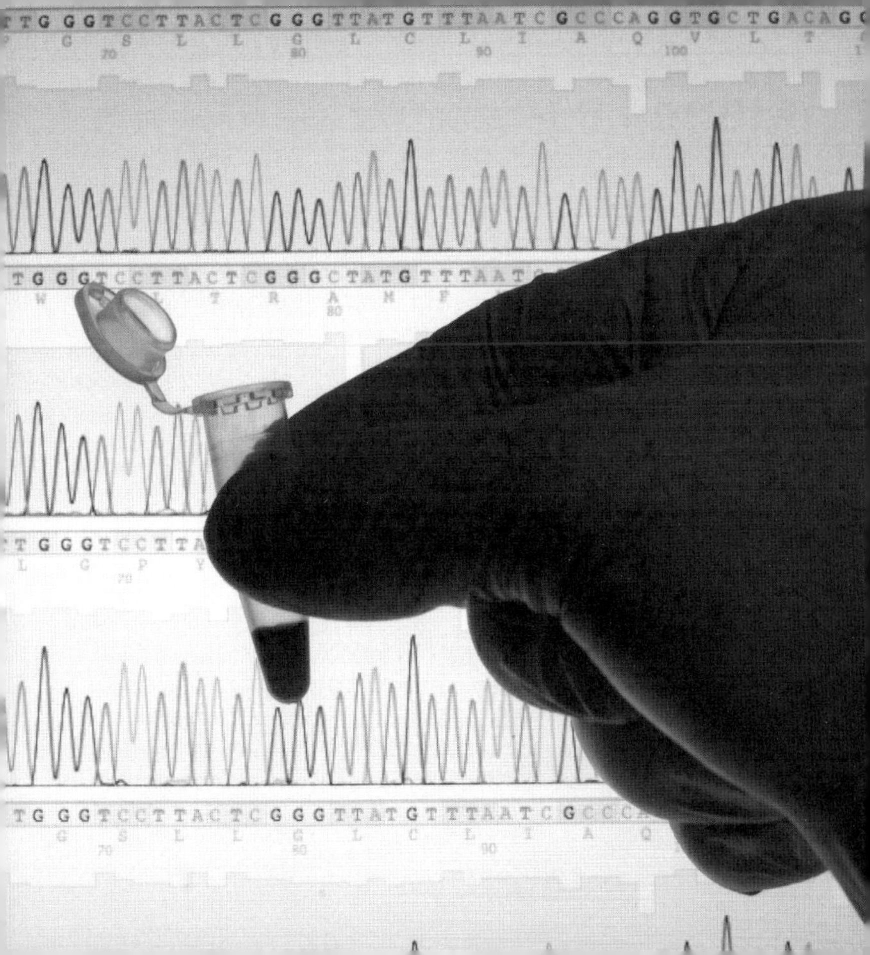

Genetic matchmaking

Women can generally smell more acutely than men. One evolutionary reason for this is to help identify toxins in food that might pass to their babies in breast milk – but it also appears to be linked to mate choice. In 1995, the Swiss biologist Claus Wedekind carried out the famous 'sweaty T-shirt' experiment. He asked male participants to wear the same shirt while sleeping for two nights running. He then asked female participants to rate the smells of each shirt. The results showed that no particular shirt was regarded as more desirable than any other, but the women tended to prefer the smells of men who had different HLA profiles to them (see page 328). The reason for this seems obvious – selecting a partner with a different HLA can ensure any offspring are better able to fight disease, and also suggests that the mate is unlikely to be a close relative. In the wake of these discoveries, companies now offer genetic profiling to couples interested in putting science before romance – though some experts have dismissed the findings as simplistic.

Gene therapy

Imagine if it were possible to replace faulty genes, or even add new ones to fight a disease. This is the goal of gene therapy, a potentially revolutionary new field of medicine that has been progressing slowly but surely since the late 1980s. There are very real dangers if something goes wrong, but signs of success are appearing.

Genes must be introduced to the body by a vector – a carrier mechanism of some sort some sort. Viruses make good vectors, but can be attacked by the immune system, and the prospect of an artificial human virus escaping into the wild is the stuff of sci-fi nightmares. Non-viral vectors, such as direct injections of DNA into the blood, have had limited success. But how should success be defined? At the very least, the DNA needs to be targeted at the tissue affected by the disease. However, the rest of the body also carries the bad gene – and so could any offspring. Germ line gene therapy therefore aims to correct the problem at source, eradicating the faulty gene from the family tree altogether.

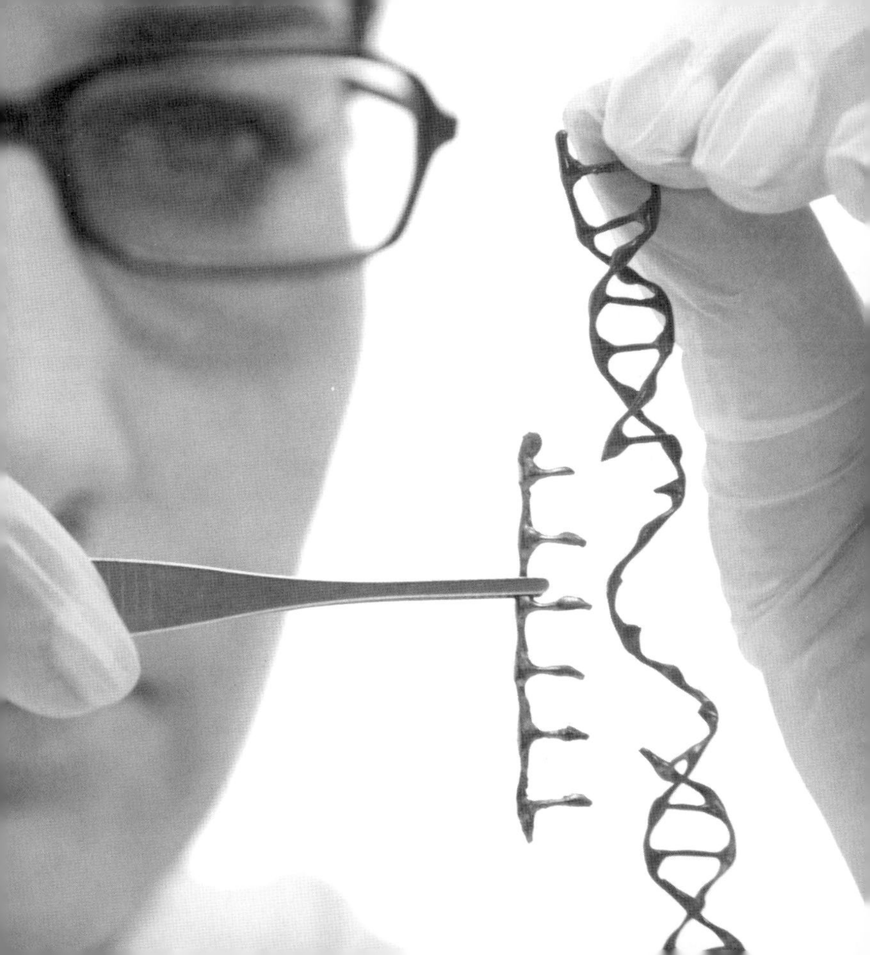

Stem cell therapy

A much-heralded future benefit of genetic technology, stem cell therapy aims to use the systems that build the body to fix otherwise incurable problems. Stem cells are the start points of a multicellular body, able to transform themselves into any cell type (see page 143). In the adult body, they also perform roles tasks such as building stomach lining and making blood cells, but once it is fully grown, most of a body's stem cells turn off. Bone marrow transplantation is a form of stem cell therapy. Healthy stem cells from a donor are put inside the bones of a patient suffering from the blood disorder leukaemia, where they replace the old marrow and restore the blood to health. The body cannot mend severe injuries to things like nerves, bones and eyes, and so researchers hope to fix these, too, with stem cells. Cells may be taken from elsewhere in the body and 'reset' so they work anywhere, or, more controversially, they can be harvested from an embyro cloned for the purpose. It is still early days, but several successes indicate that stem cell therapy will become part of everyday medicine in the future.

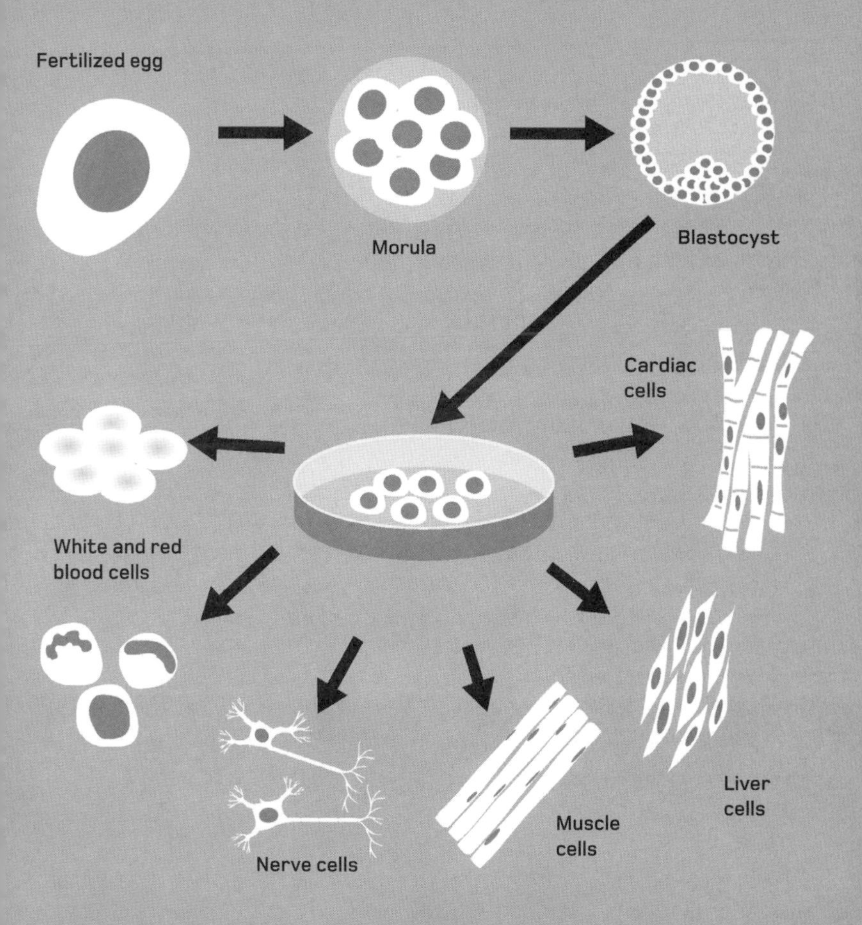
Fertilized egg
Morula
Blastocyst
Cardiac cells
White and red blood cells
Nerve cells
Muscle cells
Liver cells

Designer babies

The phrase 'designer babies' comes laden with powerful emotions, raising comparisons between babies and high-end luxury goods, such as handbags and shoes, that can be made to order. But it also suggests that genetic technology can be used to remove inherited defects that might otherwise make the child's life a misery. As with many aspects of modern genetic science, technology is once again driving an ethical debate.

While it is almost universally agreed that it is unethical to screen sperm, eggs or embryos for sex or superficial genetic traits such as hair colour, it is already possible to edit their genomes to replace disease-causing genes with healthy versions. But why stop us there? Why not ensure the best genes for intelligence and looks are included as well? As yet, such genes are not known to exist, but where should the hypothetical line be drawn? The arguments in favour of clinical intervention are hard to resist, but are we right to block those who wish to go further?

Synthetic biology

Picture a machine of the future, some device used for lifting and shifting. Is it a mechanized hunk of metal and plastic, or made of flesh and bone? We copy biological body shapes for our robots, so why not use biological materials as well – or better still, merge the two? Such a vision of the future would be the product of synthetic biology, an emerging field where scientists take what they know about genetics, cell biology and anatomy to create organisms from scratch.

In 2010, the first artificial bacterium was produced, using the cell from a pre-existing bacterium with its DNA removed and replaced with a synthetic genome written by engineers. More recently, engineers have built cell-like vesicles out of the same lipids used in cell membranes, and they are now looking at ways of creating entirely functional cells out of synthetic, non-biological materials. It may take decades rather than years, but the more we learn about the way genes, cells and bodies work, the easier it will be to make our own versions.

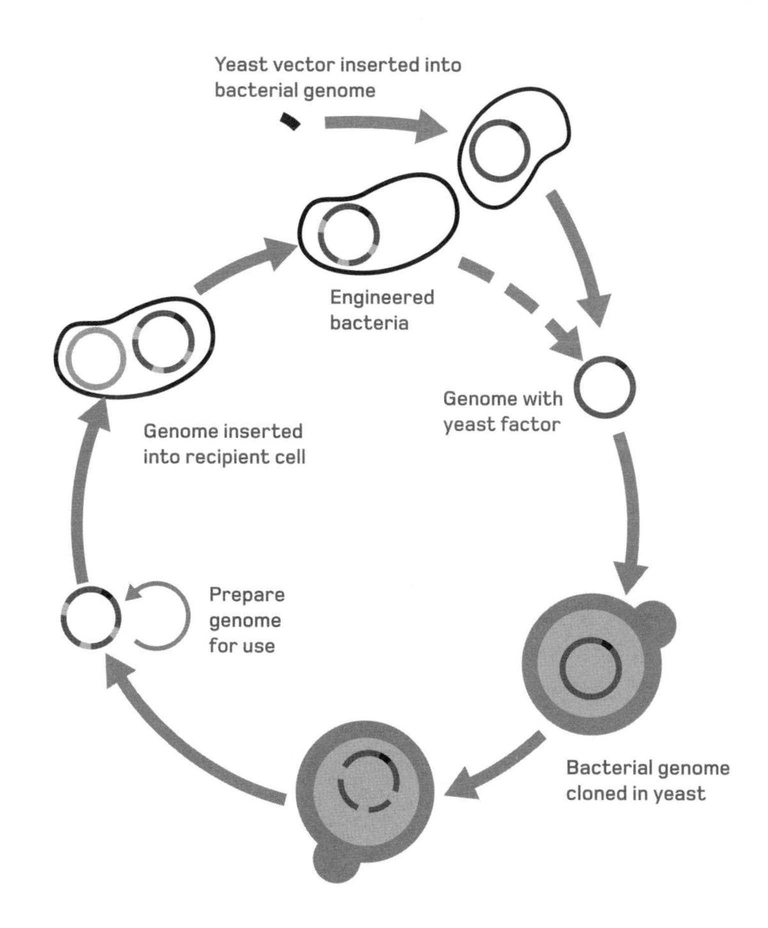

Yeast vector inserted into bacterial genome
Engineered bacteria
Genome with yeast factor
Bacterial genome cloned in yeast
Prepare genome for use
Genome inserted into recipient cell

XNA: Artificial DNA

The term XNA stands for 'xeno nucleic acids' – laboratory-made chemicals that do everything DNA and RNA can do. (*xeno* is Greek for 'other'). In 2015, researchers succeeded for the first time in using a strand of pre-programmed XNA to synthesize a protein. But why should we reinvent DNA, one of the most powerful creations in the natural world?

Synthetic nucleic acids were first produced by evolutionary biologists researching alternatives that might have competed with RNA and DNA at the dawn of life on Earth (see page 314). The next step was to build XNAs that mirrored the form and function of DNA, using the same pairings of nucleotide bases, but are much more robust in the face of chemical attacks and temperature changes. This has opened up startling possibilities: could XNA be used inside synthetic cells, perhaps creating a whole new domain of life? More immediately, gene therapy might be used to replace DNA with robust XNAs, allowing us to artificially improve our own genomes.

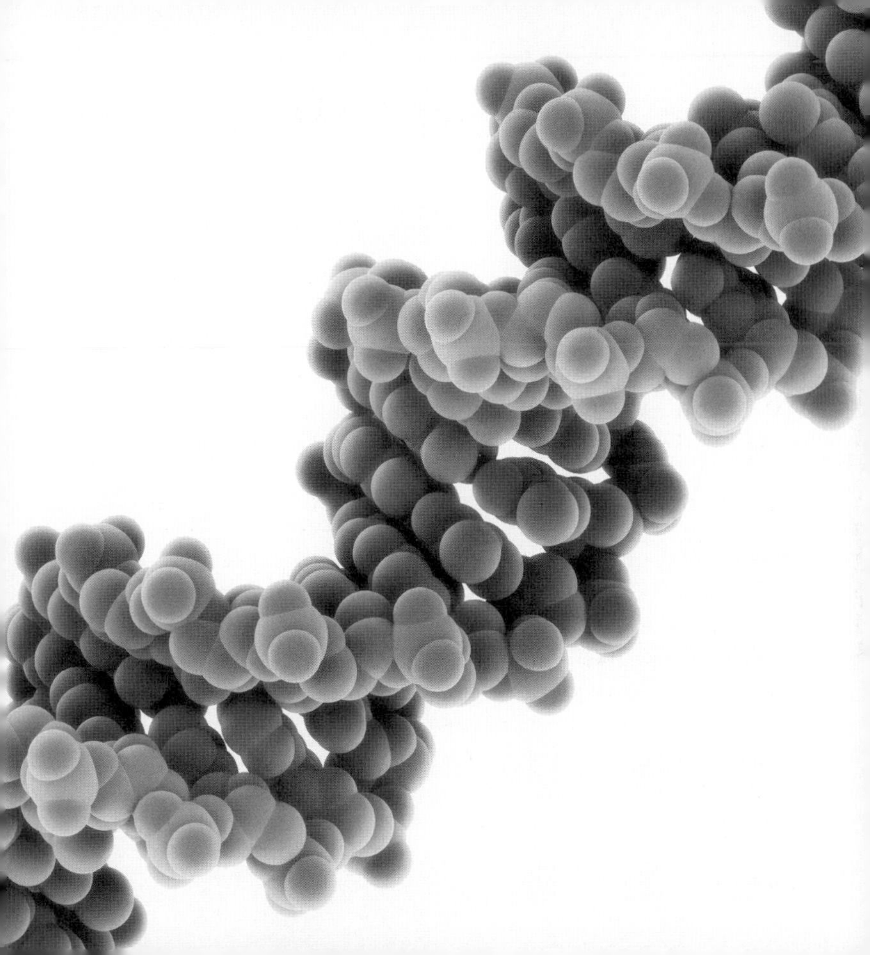

Glossary

Adaptation
A physical or behavioural trait evolved to allow an organism to survive in its environment.

Allele
A version of a gene; the gene for eye colour, for instance, has several alleles.

Amino acid
An organic compound containing nitrogen; chains of amino acids form proteins.

Cell
The smallest self-contained unit of life, from which all organisms are formed.

Chromatid
A duplicated chromosome; chromatids usually pair together, but can also act as chromosomes on their own.

Chromosome
A carrier for genetic material found in the nucleus of cells of complex organisms.

Codon
A three-molecule unit in a gene that represents an amino acid within the larger protein molecule encoded by that gene. The anticodon is the mirror image of a codon, used in genetic coding.

Diploid
Describing a cell that contains a double set of genetic material, with one set inherited from each parent.

DNA
Deoxyribonucleic acid, a ladder-like spiral molecule whose structure stores genetic information.

Endosymbiosis
A theory of how eukaryote cells evolved from smaller prokaryote cells living and working together.

Enzyme
A protein that is involved in metabolism by controlling a specific reaction needed for life.

Eukaryote
An organism with a body made from complex cells containing a nucleus and other organelles.

Evolution
The transformation of organisms over time through the interaction of outside influences and inherited traits.

Exon
A section of genetic material that carries code for a gene.

Fitness
How well an organism is suited to its environment compared to others of its species.

Gene
The unit of inheritance. It can be regarded as a strand of

DNA that codes for a certain protein, or as a distinct hereditary characteristic.

Gene pool
The total accumulation of alleles found in a population.

Genome
The full collection of genetic material of a species, including genes and non-coding DNA.

Genotype
A description of the alleles carried by an organism.

Haploid
Describing a cell that contains only a single set of genes.

Intron
The section of inherited DNA that carries no coded instructions for genes.

Mendelian
Referring to the core ideas in genetics, formulated by Gregor Mendel in the 1860s.

Mutant
An organism that carries a novel allele, or mutated gene.

Nucleotide
A nucleic acid chemical found in DNA and RNA; In DNA the nucleotides frequently form pairs, while in RNA they are single.

Nucleus
The region of a eukaryotic cell containing most of its genetic material.

Organelle
A machine-like structure in a cell that performs a particular set of functions.

Phenotype
A description of the physical and behavioural traits of an organism, as produced by a genotype.

Polymer
A long molecule made up of smaller units, or monomers, bonded together in a chain; proteins and DNA are both types of polymer.

Prokaryote
An organism with a small and simple cell that lacks organelles.

Protein
A complex molecule used by all living things to build structural body parts and muscle and as enzymes.

Respiration
The process that takes place in every living cell to extract energy from a food source, such as sugars.

Substrate
The material that is acted upon by an enzyme.

Taxonomy
The science of classifying organisms according to how they are related.

Zygote
The first cell of a living body.

Index

Quercus
New York • London

First published in the United States by Quercus in 2016

ISBN 978-1-68144-333-1

Library of Congress Control Number: 2016930913

Distributed in the United States and Canada by Hachette Book Group
1290 Avenue of the Americas
New York, NY 10104

Manufactured in China

10 9 8 7 6 5 4 3 2 1

www.quercus.com

Picture credits 9: Benjamin S Fischinger/Shutterstock; 15: Raul654/Wikimedia; 21: IndustryAndTravel/Shutterstock; 23: Eric Gevaert/Shutterstock; 25: Elnur/Shutterstock; 27: Ollyy/Shutterstock; 29: Zagorodnaya/Shutterstock; 31: gopixa/Shutterstock; 33: Wellcome Images/Wikimedia; 37: MurzikNata/Shutterstock; 41: Panaiotidi/Shutterstock; 45: Arthimedes/Shutterstock; 47: Pleple2000/Wikimedia; 51: aodaodaodaod/Shutterstock; 57: Dreamy Girl/Shutterstock; 65: NOAA Photo Library; 67: Designua/Shutterstock; 69: Designua/Shutterstock; 71: BlueRingMedia/Shutterstock; 75: Peter Gardiner/Science Photo Library; 77: Designua/Shutterstock; 79: Science & Society Picture Library; 89: World History Archive / TopFoto; 91: Richard Wheeler, richardwheeler.net; 93: diepre/Shutterstock; 105: Capreola/Shutterstock; 115: Iculig/Shutterstock; 119: TessarTheTegu/Shutterstock; 125: Designua/Shutterstock; 129: Designua/Shutterstock; 137: lotan/Shutterstock; 139: Ed Uthman, MD/Wikimedia; 145: AkeSak/Shutterstock; 147: eveleen/Shutterstock; 149: Birute Vijeikiene/Shutterstock; 151: Testudo/Wikimedia; 153: FamVeld/Shutterstock; 155: dr OX/Shutterstock; 157: s_bukley/Shutterstock; 161: MedievalRich/Wikimedia; 163: Spleines/Wikimedia; 165: Waraphan Rattanawong/Shutterstock; 167: krasky/Shutterstock; 169 t: metha1819/Shutterstock; c: Eric Isselee/Shutterstock; b: Susan Schmitz/Wikimedia; 171: Rudmer Zwerver/Shutterstock; 173: Istvan Csak/Shutterstock; 175: Dave Pusey/Shutterstock; 185: Maggy Meyer/Shutterstock; 187: Johan Swanepoel/Shutterstock; 189: GUDKOV ANDREY/Shutterstock; 191: Eric Isselee/Shutterstock; 193: JHVEPhoto/Shutterstock; 195: Olaf Leillinger/Wikimedia; 203: Federico Massa/Shutterstock; 207: Hein Nouwens/Shutterstock; 209: Mary Evans Picture Library; 213: Rama/Wikimedia; 215: aarrows/Shutterstock; 221: paula french/Shutterstock; 233: Volodymyr Goinyk/Shutterstock; 239 tl: dianewphoto/Shutterstock; tr: Galyna Andrushko/Shutterstock; bl: Tischenko Irina/Shutterstock; br: Nadezda Zavitaeva/Shutterstock; 247: seanbear/Shutterstock; 251: NOAA Photo Library; 253: Jim Peaco, National Park Service/Wikimedia; 255: irin-k/Shutterstock; 257: Jesse Nguyen/Shutterstock; 259: Rich Carey/Shutterstock; 266-7: Palenque/Shutterstock; 269: David Osborn/Shutterstock; 271: Sergei25/Shutterstock; 273: Nick Fox/Shutterstock; 275: LeonP/Shutterstock; 277: Victorian Traditions/Shutterstock; 281: Adrian Kaye/Shutterstock; 281: Katoosha/Shutterstock; 285: Chaikom/Shutterstock; 287: Brian Gratwicke/Wikimedia; 289: BMJ/Shutterstock; 291: Hindustan Times; 293: Andrey Bayda/Shutterstock; 297: Vadim Petrakov/Shutterstock; 299: Pressmaster/Shutterstock; 307: Didier Descouens/Wikimedia; 311: ESA/Rosetta/NAVCAM; 315: Richard Bizley/Science Photo Library; 319: Nicolas Primola/Shutterstock; 325: Lonely/Shutterstock; 327: InvictaHOG/Wikimedia; 329: Monkey Business Images/Shutterstock; 331: Markus Mainka/Shutterstock; 335: Denis Kuvaev/Shutterstock; 343: Designua/Shutterstock; 347: wong sze yuen/Shutterstock; 351: Alexey Losevich/Shutterstock; 353: wavebreakmedia/Shutterstock; 355: petarg/Shutterstock; 357: Hulton Archive; 359: Rido/Shutterstock; 361: Wichy/Shutterstock; 363: johnbraid/Shutterstock; 367: Anton Havelaar/Shutterstock; 369: Ross Stevenson/Shutterstock; 371: Ipatov/Shutterstock; 373: Bbski/Wikimedia; 377: oticki/Shutterstock; 379: designelements/Shutterstock; 381: Studio_G/Shutterstock; 383: Andrew M. Allport/Shutterstock; 387: Toni Barros/Wikimedia; 393: science photo/Shutterstock; 395: damerau/Shutterstock; 397: Pressmaster/Shutterstock; 399: GeK/Shutterstock; 403: Oksana Kuzmina/Shutterstock; 407: Laguna Design/Science Photo Library;

All other illustrations by Tim Brown